SURFACE TREATMENT AND FINISHING OF ALUMINIUM

OTHER TITLES IN THE SERIES

ALLSOP & KENNEDY	Pressure Diecasting, Part 2
ANTHONY	Engineering Metrology
BYE	Portland Cement
DAVIES	Protection of Industrial Power Systems
HENLEY	Anodic Oxidation of Aluminium and Its Alloys
HOLLAND	Microcomputers for Process Control
LANSDOWN	Lubrication
LANSDOWN & PRICE	Materials to Resist Wear
MASKALL & WHITE	Vitreous Enamelling
MORGAN	Tinplate and Modern Canmaking Technology
NEMENYI	Controlled Atmospheres for Heat Treatment
PARRISH & HARPER	Production Gas Carburising
SHEWARD	High Temperature Brazing in Controlled Atmospheres
UPTON	Pressure Diecasting, Part 1
WILLIAMS	Troubleshooting on Microprocessor Based Systems

SURFACE TREATMENT AND FINISHING OF ALUMINIUM

R. G. KING
British Alcan Aluminium plc, Gerrards Cross, UK

PERGAMON PRESS
OXFORD · NEW YORK · BEIJING · FRANKFURT
SÃO PAULO · SYDNEY · TOKYO · TORONTO

U.K.	Pergamon Press, Headington Hill Hall, Oxford OX3 0BW, England
U.S.A.	Pergamon Press, Maxwell House, Fairview Park. Elmsford, New York 10523, U.S.A.
PEOPLE'S REPUBLIC OF CHINA	Pergamon Press, Room 4037, Qianmen Hotel, Beijing, People's Republic of China
FEDERAL REPUBLIC OF GERMANY	Pergamon Press, Hammerweg 6, D-6242 Kronberg, Federal Republic of Germany
BRAZIL	Pergamon Editora, Rua Eça de Queiros, 346, CEP 04011, Paraiso, São Paulo, Brazil
AUSTRALIA	Pergamon Press Australia, P.O. Box 544, Potts Point, N.S.W. 2011, Australia
JAPAN	Pergamon Press, 8th Floor, Matsuoka Central Building, 1-7-1 Nishishinjuku, Shinjuku-ku, Tokyo 160, Japan
CANADA	Pergamon Press Canada, Suite No. 271, 253 College Street, Toronto, Ontario, Canada M5T 1R5

First edition 1988

Library of Congress Cataloging in Publication Data
King, R. G.
Surface treatment and finishing of aluminium.
(The Pergamon materials engineering practice series)
1. Aluminum—Finishing. I. Title. II. Series
TS555.K46 1988 673'.7227 87-16070

British Library Cataloguing in Publication Data
King, R. G.
Surface treatment and finishing of aluminium.—(The Pergamon materials engineering practice series).
1. Aluminium—Finishing
I. Title
673'.7227 TS655

ISBN 0-08-031137-7 (Hard cover)
ISBN 0-08-031138-5 (Flexicover)

Printed in Great Britain by
Hazell Watson & Viney Limited
Member of BPCC plc
Aylesbury Bucks

To Catriona

Materials Engineering Practice

Foreword

The title of this series of books "Materials Engineering Practice" is well chosen since it brings to our attention that in an era where science, technology and engineering condition our material standards of living, the effectiveness of practical skills in translating concepts and designs from the imagination or drawing board to commercial reality, is the ultimate test by which an industrial economy succeeds.

The economic wealth of this country is based principally upon the transformation and manipulation of *materials* through *engineering practice*. Every material, metals and their alloys and the vast range of ceramics and polymers, has characteristics which require specialist knowledge to get the best out of them in practice, and this series is intended to offer a distillation of the best practices based on increasing understanding of the subtleties of material properties and behaviour and on improving experience internationally. Thus the series covers, or will cover, such diverse areas of practical interest as surface treatments, joining methods, process practices, inspection techniques and many other features concerned with materials engineering.

It is to be hoped that the reader will use this book as the base on which to develop his own excellence and perhaps his own practices as a result of his experience and that these personal developments will find their way into later editions for future readers. In past years it may well have been true that if a man made a better mousetrap the world would beat a path to his door. Today, however, to make a better mousetrap requires more direct communication between those who know how to make the better mousetrap and those who wish to know. Hopefully this series will make its contribution towards improving these exchanges.

MONTY FINNISTON

Contents

Introduction

Anodising of aluminium and its alloys is an important aspect of finishing aluminium, and the subject has been well covered in the book *Anodic Oxidation of Aluminium and its Alloys* by Henley (Pergamon, 1982).

The aim of this book is to describe the various other surface treatments and finishes available to and being used in industry. It is hoped that it will be a guide to those who are already involved in the processing of aluminium, or those who are intending to enter the industry. Also it is hoped that the book will help those who are considering using aluminium, such as designers and specifiers, to show what finishes are available, and to simplify the choice of the most suitable method.

The alternative methods for the surface treatment of aluminium, which are carried out for one or other of several reasons, are covered in this volume. The metal is basically very durable and can be used in many instances in the unfinished state. But sometimes protection also has to be given when it is used in special environments, or a long-lasting decorative finish is required, or one may wish to alter the surface to provide different properties. There are quite a few treatments available, and while some of the finishing processes can be carried out on other metals, there are certain features which have to be taken into account when coating aluminium.

Many of the processes described are not new, although continued improvements and developments are taking place. One particular change is in the introduction of new methods of organic coating, where powder coating and electropainting of architectural aluminium has expanded rapidly, partly at the expense of anodising and conventional painting. Improvements are also being made in paints themselves, which can only be of benefit in the production of coloured finishes on aluminium.

Engineering coatings requiring surface hardness and wear resistance are important; they are in a specialised field and although perhaps not so widely carried out, they fulfil a particular role.

In the light of present day economics, perhaps undue stress has been laid on the cost of a product. Aluminium compared with steel can be expensive, due to dependence on electrical energy during its production. Its great advantage lies in the many instances where it can be used in the unfinished state. Metal finishing can be an expensive process, and justification is required to carry it out. Where aluminium gains is that the finishes applied to aluminium are more durable than on other metals, and therefore the increased life expectancy of finished aluminium goods is a beneficial reward.

As indicated in Figure 0.1, two of the largest markets for aluminium are

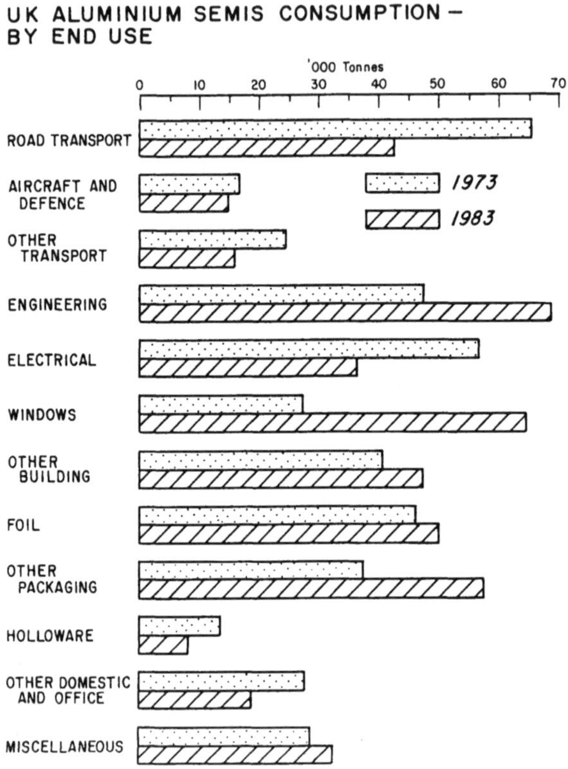

FIGURE 0.1.

transport and architecture. The first is largely functional with a shorter life-span, and decorative finishing is of less importance. Architectural aluminium, although often equally functional, has also an aesthetic role to play, and long life expectancy. Hence finishing plays an important part, and one has a choice in selecting one of the durable organic coatings now available, or the even longer lasting but more expensive anodised finishes.

Functional coatings include the resistant coatings required for lining food containers, wear resistance of electroplated or electroless nickel coatings or the corrosion resistance of special paints.

The selection of a particular process depends on a number of interwoven factors such as properties of the finish, cost, availability and also health and safety aspects, all of which must be considered before reaching a decision. Information contained in this volume should provide the answer.

In one area in particular, a number of alternative finishes are available. This is in organic coatings where there are a number of different coatings which can be used for decorating aluminium architecture. The various coatings are described in Chapters 7 to 10, and a summary of their advantages is given in Chapter 11 to enable a selection to be made.

Chapter 1

Mechanical and Textured Finishes

1.1 INTRODUCTION

One of the features of aluminium is the number of ways in which the surface of the metal can be physically altered or modified to produce different textures to decorate it or to mask any surface imperfections that may have arisen during fabrication. In this chapter, mechanical methods will be described, in contrast to chemical or electrochemical methods such as etching or brightening in acid or alkali solutions.

1.2 FINISHES ON ROLLED PRODUCTS

Whilst most finishes can be applied to the fabricated article, in some cases it is possible to produce aluminium sheet with a special textured or patterned finish during the rolling stages.

When aluminium is rolled to produce sheet or coil, the surface depends on the texture of the steel rolls. Normally, for ease of production, the appearance of aluminium is dull and grey corresponding to the surface finish of the rolls which are lightly ground to allow them to grip the aluminium and permit higher rolling speeds. The surface of the aluminium sheet may also be slightly banded in the direction of rolling as a result of variations in texture of the metal. This is the typical surface of aluminium sheet used for non-critical applications where appearance is not important, such as in general fabrication and general engineering or where special finishes are to be applied later.

It is also possible to produce sheet with a bright mirror-like surface, by using rolls with polished surfaces, and this product has certain applications such as name plates and other decorative articles.

One disadvantage of bright rolled finishes is the expense of production, owing to the need of special manufacturing techniques, such as low rolling speeds; often on hand mills for individual sheets rather than by the more usual coil or strip rolling. Furthermore the rolled surface is vulnerable to mechanical damage, requiring temporary protection during transit and subsequent fabrication. In use it must be protected by lacquering or anodising to avoid dulling or damage from scuff marks.

1.2.1 Patterned Sheet

A variety of patterns, usually of shallow designs can be applied to sheet, sometimes from patterned rolls but often by passing two sheets of aluminium through a hand mill with a steel template between the aluminium sheets (Figure 1.1). The designs in relief make the sheet suitable for many decorative uses, such as fascia panels for domestic articles, for example radios, washing machines, etc. As well as providing decoration, the patterns help to mask any surface variations in the rolled surface, and certain patterns are helpful in obscuring any scratching that can take place in service, making the sheet useful in situations where wear may take place. Deeper and more sharply defined patterns can be applied by stamping sheet with a metal die to resemble "engine-turning". This process has been used for decorating finished articles, such as cigarette cases.

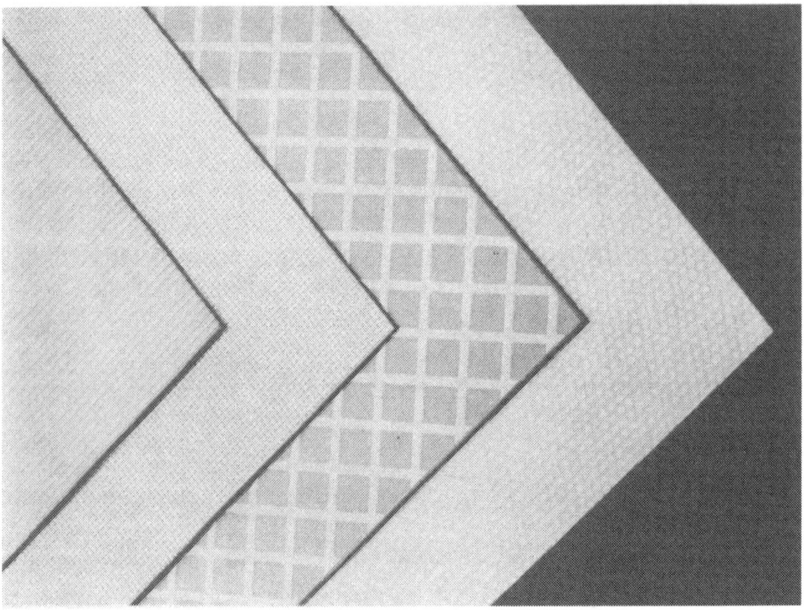

FIGURE 1.1. Patterns produced on aluminium sheet during rolling – British Alcan Aluminium plc

Patterns, such as dimpling of the surface have been produced on components such as light reflectors formed by metal spinning, not only for decoration, but for breaking up the surface to produce a more diffuse reflection. This is produced by using a patterned former during the spinning process.

So-called "Stucco" finish is available for use on building sheet, where patterned rolls are employed to break-up the surface with random and shallow indentations. This has the benefit of reducing possible glare and reflections on freshly rolled sheet which might cause problems on a newly erected building.

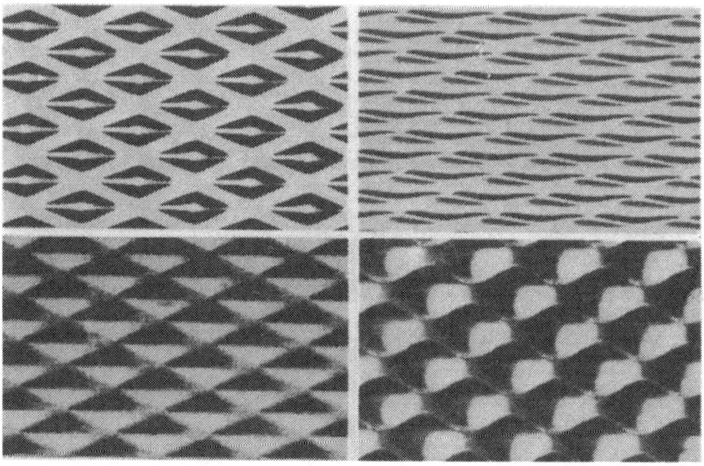

FIGURE 1.2. Some rigidised patterns for aluminium – Rigidised Metals Ltd

Some raised three-dimensional patterns are produced by rolling heavy gauge sheet used as tread plate on stairs and gangways where the projecting lozenges have the functional effect of producing a non-skid surface.

One particular form of patterning is known as "Rigidising"* which is a rolling process, using male and female patterned rolls to provide a three-dimensional pattern in the form of dimples (Figure 1.2). The formation of these dimples has a slight work hardening effect on the aluminium and the stiffening effect of the upstanding pattern makes the sheet more rigid. Sheet processed in this way has been used for making containers where the increased rigidity is of benefit. Rigidising can be carried out on sheet pre-coated with a paint that is flexible enough to be deformed.

1.2.2 "Butler Finish"

A fine, textured finish with the appearance of fine linishing, can be produced by using a linished roll for the final rolling operation. Again, the appearance of the finished product depends on the uniformity of the grind-

* Rigidised Metals Ltd., Aden Rd, Ponders End, Enfield, Middlesex.

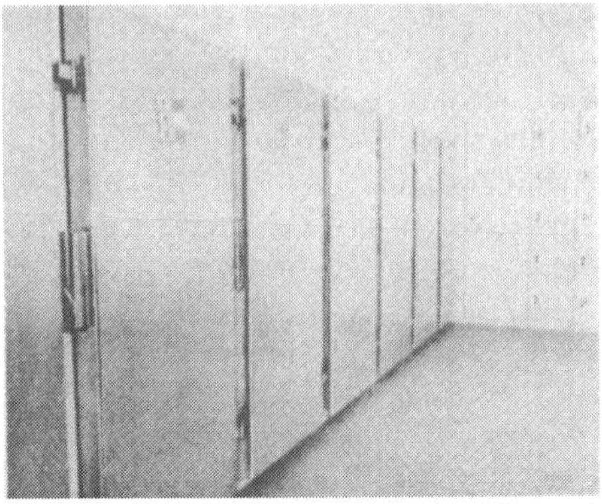

FIGURE 1.3. Rigidised panels used for door panels of Leeds City Mortuary

ing operation on the steel rolls and on other factors, and high standards are required during manufacture to provide this uniformity.

1.3 APPLIED MECHANICAL FINISHES

The finishes so far described are usually those that are supplied on sheet provided by the manufacturer.

Probably of more concern to the fabricator, is the variety of ways in which aluminium can be processed by applied mechanical methods, either to remove any blemishes that might have taken place during manufacture, or to provide special textured effects. These can range from a coarse, directional effect produced from linishing or grinding (see 1.3.1) which involves metal removal, down to a uniform and bright polished finish resulting more in metal flow at the surface, rather than actual metal removal.

The operations can be carried out manually, with the advantage of flexibility and closer control, but depend on operator skill and are more labour intensive. For large throughputs, automation can be beneficial with higher speeds and more consistent results, but with higher capital costs.

There are various stages that may need to be followed in obtaining a bright, reflected and polished finish on an aluminium article. If the original surface is uneven, badly marked or mechanically damaged, the complete sequence has to be followed, but if the aluminium is in reasonable condition, the first stages can be omitted.

1.3.1 Linishing or Grinding

This is an operation to remove surface metal by means of a coarse abrasive such as emery, aluminium oxide or carborundum (silicon carbide) mounted with abrasives on a moving belt or a rotating wheel. (Grinding wheels or

FIGURE 1.4. Abrasive belt grinding of aluminium extrusions – Greif Machines (UK) Ltd

stones should not be used on aluminium as they become clogged up by the soft metal particles.)

Work to be processed is held against a driven soft wheel over which passes the abrasive belt, the other end of the belt being held in tension by a free-running wheel, the set-up being known as a back-stand idler.

Use of abrasive belts will remove metal rapidly and provide a larger contact area than coated wheels described later. They should be lubricated with tallow or other suitable grease or lubricant to prolong the life of the belt and provide a more even finish.

Coarse abrasives are used mainly for grinding, in particular, on castings to remove the effects of risers and runners after they have been sawn off, and to smooth over the surfaces. Linishing has also been used to remove the residues from "flash" after flash-butt welding.

Another application in the use of linishing is the pretreatment prior to buffing or polishing of fabrications such as aluminium bumpers where the forming processes may have damaged the surface, and to remove excess weld metal.

Linishing has many applications in the field of surface finishing where a uniform textured finish can readily be produced. Whilst it is difficult to specify the particular finish required, attempts have been made to do so by selection of the grade of the abrasive used, such as proposed by the Aluminium Association of America*:

Coarse finish	80–100 mesh aluminium oxide grit
Medium	180–220 mesh aluminium oxide grit
Fine	320–400 mesh aluminium oxide grit.

The recommended speed of the linishing belt is 6000 ft/min.

Belt linishing is mainly used for flat surfaces, and whilst small objects can be treated by hand, long lengths of extrusions are more economically processed by machines with one or more belts which can treat several faces at one pass.

Linishing of extrusions required for architectural anodising has been suggested as an alternative to deep chemical etching to produce a matt surface, provided the directional effect can be tolerated. Some chemical etching is still required to clean a linished surface if required as a pretreatment before anodising.

Linishing by mechanical means overcomes some of the disadvantages of chemical etching in that it reduces liquid effluent problems, is a more effective method of overcoming surface imperfections such as die-lines and can eliminate occasional problems of atmospheric surface corrosion which can appear as pitting after chemical etching.

Linishing of flat sheet is carried out on machines using wide belts, and the linear texture produced has been applied to sheet in the medium strength alloy 6082 without further treatment when used as infill panels on escalators, etc. Care is needed to provide a uniform appearance over the width of the sheet.

Whilst belt linishing is ideal for flat surfaces on sheet or on suitable surfaces of extrusions, it is not easy, or indeed possible, to obtain a uniform finish on curved surfaces. An alternative to belts is the use of flexible wheels mounted on a polishing spindle on to which abrasive compounds are applied with a suitable glue. These are more gentle acting than belts and can be used on curved surfaces to provide textured finishes.

A particular case is centreless grinding of round tube. This is an arrangement of linishing belts or abrasive wheels that enables one to scurf or linish the tube in a concentric linishing direction to provide a decorative uniform surface and to remove any surface imperfections. It has been used in the treatment of aluminium vacuum cleaner tubes.

A useful method of grinding or linishing extrusions is use of a spindle mounted "flap-wheel", consisting of radial strips of emery cloth or paper. This allows some flexibility for abrading shaped surfaces.

A method of producing scratch-brushed finishes is with special wheels

* Designation system for aluminum finishes – The Aluminum Association.

consisting of an open spun texture, usually of nylon, in which is incorporated abrasives such as aluminium or silicon carbide. They are manufactured under the trade names of "Scotchbrite"* and "Vileda"† and can be produced in a variety of widths and diameters. In operation, only one pass may be necessary, and it is important to apply low pressures and low speeds (6 m/min) to avoid overheating which would melt the nylon. Although metal removal is less, dust formation is reduced. Use of these wheels is effective in producing textured finishes.

1.3.2 Polishing

The terminology of producing a polished surface on to aluminium by mechanical means, can be confusing. Polishing is generally understood to include all the stages of roughing, cutting and buffing to colouring.

The operations are carried out on spindle mounted wheels made of layers of calico or cotton of different types, to which are applied polishing compounds consisting of fine abrasives mounted in a grease or lubricant. The initial stages of polishing abrade the aluminium surface and removes metal, overcoming die-lines and other surface imperfections. The final stages of buffing and colouring operate by burnishing, resulting in flowing the metal surface to a smooth finish rather than by abrasion with metal removal.

The mops or wheels used are made for the particular operation to be used. For the initial metal removal operations, stitched mops consisting of discs of cotton or calico are sewn together providing a hard wheel on to which the polishing compounds are applied. It should be realised that polishing is done by the compound and not by the wheel.

Because of the rougher surface finish, castings may require linishing before further polishing to provide a smooth surface, but wrought products formed from sheet or extrusions should only require polishing and buffing. In many instances, components made from sheet only require buffing.

Buffing is the final stage in producing a mechanically polished surface. It is carried out on softer mops, often made of unstitched cotton discs, using polishing compounds containing a finer abrasive than used for polishing.

The field of polishing and buffing is specialised, and advice should be taken from the manufacturers of equipment.

Automation is desirable if the production rate is high enough, but the equipment is complicated and may not be flexible enough to treat a wide range of products (Figure 1.5). It is well worthwhile to take steps to reduce the necessity to polish such as by use of temporary protective coatings applied to aluminium before fabrication, and to minimise unnecessary damage through mishandling.

* 3 M Co., P.O. Box 1, Bracknell, Berkshire.

† Vileda Ltd., P.O. Box 3, Greelland, Halifax, Yorkshire.

FIGURE 1.5. Rotary table plant for polishing flat iron soles—Greif Machines (UK) Ltd

Aluminium, tending to be softer than other metals, is more easily scratched and foreign particles can be pressed into its surface which would be revealed in subsequent chemical operations where the metal surface is dissolved. The alloy to be treated is important. Metal with a hard surface, such as pure aluminium in the H8 or fully work hardened condition, is more easily polished than in the fully annealed condition where the surface can be more easily marked by drag from the polishing mop. Equally, heat treatable alloys are most easily polished in the aged or TF condition.

1.4 POLISHING

As well as being dirty, polishing is an expensive process by whatever means are used. When done manually on spindles, output can be slow, consequently, labour costs are high. Automatic polishing machines reduce labour costs, but capital costs of the equipment are high and can only be justified for high output requirements.

1.4.1 Barrel Polishing

A useful means of treating small parts is by tumbling or rotating the components in a barrel containing water and a suitable grinding medium.

The barrel can be made of wood or metal and is lined with rubber. The polishing or grinding media consists of pebbles or composite chips, and the barrel is rotated about a horizontal axis or at an angle slightly to the vertical if open ended. The barrel rotates at a low speed allowing the components to tumble over each other and over the media, when grinding and polishing of the surface takes place.

Components are loaded into the barrel along with the grinding medium, usually now ceramic chips and some water which is neutral. The barrel is rotated at the prescribed rate and time, usually at not more than 40 rpm for up to 24 hours. Unloading is carried out by tilting the contents of the barrel on to a mesh or grid, separating out the components from the polishing media.

The process has limited application for aluminium; the components must be strong or rigid enough to withstand the tumbling action, and any sharp edges become rounded during the processing. In practice, barrelling has been restricted mainly to treatment of castings, where the rough as-cast surface can be smoothed and the effects of linishing after sawing are removed. The finish produced is smooth rather than bright, and barrelling is sometimes used preliminary to conventional mechanical polishing.

Typical examples of components processed by barrelling include:

Cast door furniture
Brackets
Car seat belt buckles
Car and motorcycle components

1.4.2 Vibratory Finishing

A variant on barrel polishing is the finishing of components in a container which is agitated with a small amplitude at high speed. The process is carried out dry and small shaped ceramic chips in special shapes are mixed with the components to be treated. The process is less liable to damage the aluminium surface than barrelling, and a finer, brighter and more reflective finish can be achieved without the risk of dulling by oxidisation which can occur when drying wet components. If the process time is extended long enough, a finish suitable for further processing such as chemical brightening and anodising can be achieved.

Generally, the process is faster than barrelling, and a smoother finish can be obtained. It has the similar disadvantage that small, recessed areas may not be accessible to the polishing chips, and sharp edges cannot be maintained. Nevertheless, it is useful for treating quantities of small components, and has been used for producing a polished finish on large components up to 700 millimetres in diameter.

1.4.3 High Energy Immersion Polishing

Sometimes known as plunge polishing, a method has been developed where components are firmly fixed on a jig and immersed in a bed of abrasive particles contained in a cylinder which is revolved rapidly about a vertical axis. The jig is held rigidly, whilst the abrasive particles are energised by centrifugal forces and bring about a rapid polishing and surface grinding action.

The jig may be orientated in the polishing media to ensure all sides of the components are evenly treated.

Faster than vibratory processes or barrelling, a fine matt finish is achieved in one operation, although further stages can be introduced for finer finishes. An advantage of this method of polishing is the short processing time required, measured in seconds, allowing high rates of output.

1.5 ABRASIVE BLASTING

This is a process consisting of impinging solid particles in a stream of air or liquid on to a surface. There are two particular fields of interest to the metal finisher, these being:

(a) to remove all paint or surface oxide from the aluminium surface to provide a clean surface for subsequent treatments.
(b) to provide a uniform, satin finish for the reduction of glare or reflections.
(c) to reduce surface porosity of castings to allow further processing such as painting or anodising.

The process may also have certain metallurgical advantages such as increasing fatigue life and increasing resistance to stress corrosion. Various methods are employed.

1.5.1 Sand or Shot Blasting

This in the simplest form consists of blowing abrasive particles with compressed air through a nozzle to clean the surface, whilst at the same time imparting a roughened, matt finish. The particles can be silicon, steel shot or other hard material. If steel is used, it should be stainless since if mild steel particles are employed, any steel left embedded in the aluminium surface could be a corrosion hazard, either developing rust stains on oxidisation or providing a source of galvanic corrosion of the aluminium.

The surface provided is fairly coarse and uneven, depending on the operating conditions, but can help to provide a mechanical key for subsequent painting, although not sufficient on its own to promote adhesion.

The process has been used to provide a glare-free finish to aluminium used on road barriers avoiding dazzle and reflections from car headlights. It has also been used to renovate architectural aluminium by removing corrosion products and cleaning the surface of contaminants especially where original protection was absent or inadequate. It provides a suitable surface on which a system of protective coatings can be made to adhere.

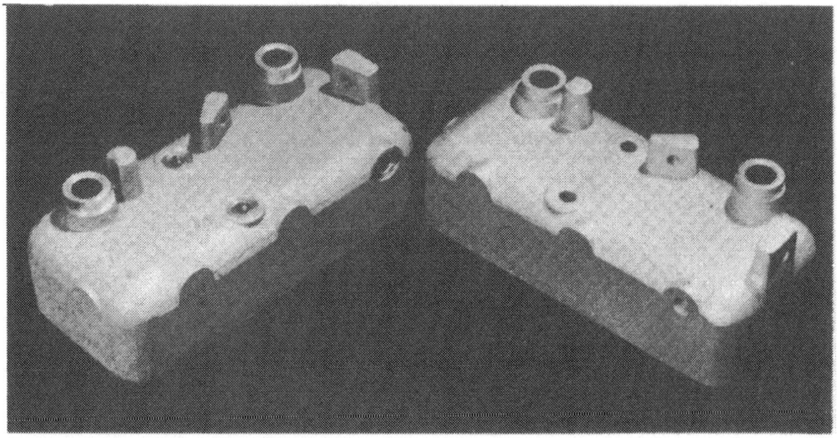

FIGURE 1.6 Aluminium castings deburred by blasting with glass beads. This blends in any defects without altering the dimensional accuracy—Vacu-Blast Ltd

More sophisticated methods can be used with a gentler action that can provide a uniform clean surface on aluminium with a smoother finish than by sand or grit blasting. In these, fine abrasives such as alumina or small glass beads are propelled on to the metal in a stream of water or air, usually in a dust-proof cabinet where the abrasive material can be re-used. (Figure 1.6)

These methods are effective in cleaning the aluminium surface, and in particular, in renovating surfaces that may have deteriorated in service through lack of maintenance or protection. Any soil or other products of oxidation can be removed to leave a clean surface before protective painting. In some cases, by use of adapted equipment, it is possible to carry out on-site treatment, where the nozzle for the abrasive is surrounded by an outer tube under vacuum to capture the spent abrasive and debris. Flexible bristles at the end provide a seal to allow collection of the particles. A disadvantage of abrasive blasting is its tendency to wrap or buckle objects made from thin gauge sheet, but it is a useful tool in cleaning, pretreatment or decoration of aluminium.

1.5.2 Shot Peening

This is a further variation of mechanical finishing, where rounded metal balls are impinged on to aluminium. It has sometimes been used on castings where the particles have a hammering effect on the surface and help to reduce the tendency for surface porosity in castings, allowing a sounder base for subsequent painting.

1.6 POST TREATMENTS

Nearly all mechanical operations on aluminium are carried out as a first step prior to subsequent treatments. Anodising is one method, but a variety of organic finishes, such as lacquering or painting is available for its preservation.

Chapter 2

Cleaning

2.1 INTRODUCTION

Compared to many other metals, the condition in which aluminium is received for manufacture of finished components is usually clean and free from many surface contaminants. In the rolled form, sheet or coil is largely free from grease or oil, since the lubricants used during the rolling stages are generally based on paraffin (kerosene) and are volatilised off during the final annealing or heat-treatment stages, or by natural evaporation. Nevertheless, the rolled surface may still be contaminated to a certain degree by oxides or airborne dirt impacted on to the surface during the rolling stages which may be sufficient to interfere with subsequent surface treatments.

Usually, therefore, a specific cleaning process must be carried out to provide a uniform and chemically active surface. Certain alloyed sheet, such as those containing copper for high-strength applications used in the aircraft industry, may have been given a protective coating of grease or lanolin to eliminate the possibility of surface corrosion during transit or fabrication, and these protectives must be removed before the final permanent surface treatments are applied.

Extrusions are generally free from grease, but the strong alloys, usually containing copper, may have been coated with lanolin-based temporary protectants by the producer which must be removed before final finishing treatments. Extruded aluminium may be contaminated with oil-based cutting fluids used during sawing to length.

The surfaces can sometimes be affected by pick-up of graphite when the extruded lengths are passed down the extrusion bench which must be removed if further surface coatings are to be applied. After machining operations, such as drilling and sawing during fabrication, further contaminants may be introduced. Drawn tubes may be coated with oil from the drawing operations, but since they are generally heat treated after drawing, these oils will have been removed before heat treatment. Sand castings usually have their surface contaminated with sand from the moulds, and there are usually present oxides etc. formed during casting or during subsequent heat treatment. Fettled surfaces may be contaminated by grease used on linishing belts.

During further manufacturing operations, such as spinning, drawing, pressing, cutting or machining, grease or oil is introduced on to the surface of components and must be removed before despatch.

If further surface treatments are to be carried out, in addition to the physical removal of oil etc., the surfaces must be chemically cleaned to remove oxides present on the surface by controlled etching of the surface which will provide a uniform base.

Depending on the situation, a variety of methods are available, some of which are common to other metals, but in some cases there are certain variations.

The method chosen will depend on a number of variants, these being:

(a) Type of soil or contaminant
(b) End use of component, including subsequent processing
(c) Cost
(d) Health and safety.

2.2 SOLVENT CLEANING

The simplest, but seldom the most effective, way of removing grease or oil, is by wiping with a clean rag, moistened with an organic solvent. This method is commonly used for on-site cleaning and where the requirements for cleaning are not critical. Its efficacy depends on a plentiful supply of clean rag and solvent. Such solvents are expensive which prevents one from simply washing over or swabbing the surface with them, a method which would also be a hazard to health, possibly a fire risk, and cause pollution. It is usual, therefore, to moisten a cloth with solvent and remove gross contamination by wiping, using a fresh area of cloth frequently and a minimum amount of solvent. The disadvantage is that it is impossible to remove all traces of contaminant, as the action will tend to smear the surface and leave staining. Organic solvents are volatile and to a greater or lesser extent give off toxic fumes, which may also be inflammable. Protective gloves must be worn to avoid contact with the skin. The vapour of certain solvents may in themselves be relatively harmless, but may break down if passed through a lit cigarette to form poisonous by-products. Smoking must, therefore, be avoided during the operation. It is essential, also, that adequate ventilation is available to remove any fumes or vapours. Certain solvents, at one time not considered to be toxic, may now be suspected of being carcinogenic and are considered dangerous on that account. Trichloroethylene, for instance, is limited by many authorities to a maximum concentration of 100 ppm in the atmosphere.

Benzene is one such solvent which must never be used, although at one time it was freely available for household cleaning purposes.

The following solvents are used for degreasing by wiping:

Paraffin (Kerosene) This is useful for on-site cleaning grease, oil or lubricants. It will evaporate, leaving no deposits and, although inflammable, has a comparatively high flash point.

White Spirit Similar in use and performance to paraffin, it is used for cleaning down stove painted aluminium architectural sheet products prior to repainting or for touching up of small areas after accidental damage.

Carbon Tetrachloride This has the advantage of being non-inflammable, and is a useful solvent for many greases and oils.

Trichlorethane This is one of several organic solvents of its type that can be safely used for hand wiping. It is available under trade names such as Chlorothene NU* or Genklene† and supersedes earlier use of trichlorethylene which is now considered to be a possible health hazard when used in this way.

Methylated Spirits has the advantage of being cheap and efficient in removing fingerprint marks, provided they are treated early enough before any formation of corrosion stains. Whilst being volatile, it has also the advantage of being miscible with water.

2.3 EMULSION CLEANING

Emulsion cleaning consists of immersing components in a bath containing organic solvents and wetting agents added to a paraffin base. The aluminium is soaked in the mixture for 5–10 minutes, allowing grease or oil to be loosened or removed, followed by one or two running water rinses to emulsify and remove any residual surface contaminants.

Although this process is now little used, it is particularly successful as a soak cleaner in softening and removing residues of polishing compound left on aluminium, particularly when the compound might have become dried or hardened on the surface. It is cheap to use, being operated at room temperature, but problems can arise over treating emulsified effluent water. It can also leave an oily film on the surface requiring a further chemical cleaning stage to remove these residues. A subsequent chemical cleaning process may not be necessary before chemical brightening or alkaline electrochemical brightening since these solutions are themselves capable of removing any remaining residues from the cleaning process.

* Dow Chemical Co.
† ICI Ltd.

2.4 VAPOUR DEGREASING

Aluminium components can be degreased by suspension in a vessel containing the hot vapour from boiling organic solvents held in the bottom

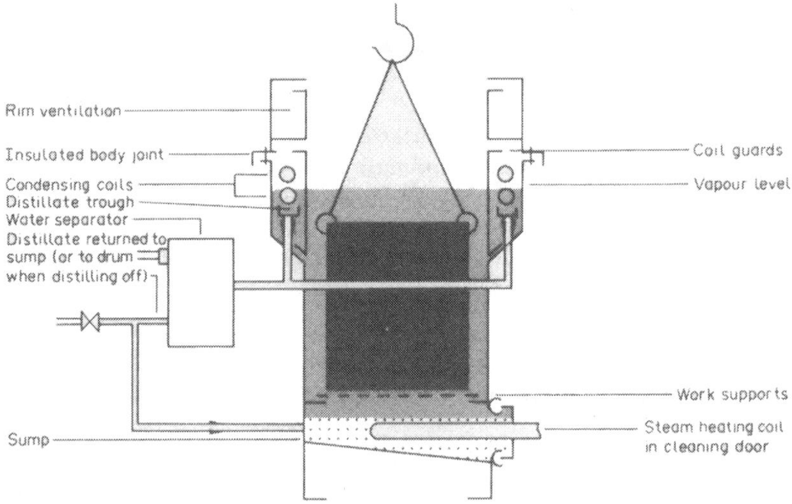

FIGURE 2.1. Single compartment vapour type degreasing plant — ICI Ltd

of the tank. The cleaning action occurs mainly when the hot vapour condenses on aluminium which is cold when it is first immersed in the vapour, and the liquid falls down to the bottom of the tank taking with it oil or grease contaminants. The advantages are that the vapour and condensed solvents are hot, making the solution of greases or oils more effective, and that the vapour is always clean, being continually distilled from the boiling solvents at the base of the vessel. Vapour degreasing, whilst removing most soils and greases, does not leave a chemically clean surface, and is usually followed by chemical cleaning if further processes are required. It is a useful method of cleaning oil or grease from small parts, which can be degreased in bulk by loading into metal baskets. (Figures 2.1 and 2.2)

The tank is made from galvanised steel with a sloping bottom to facilitate draining of contaminated solvent and cleaning. The solvent is heated to boiling point by thermostatically controlled heaters using steam or hot water tubes or electric elements. Gas immersion heaters have also been used, but are not recommended.

The hot vapour from the boiling solution fills the space above the liquid level, and is contained in the tank by circulating water cooling tubes at the top which condenses the upper level of vapour. Lip extraction at the top of the tank is also necessary to prevent escape of solvent vapour into

FIGURE 2.2. Plant for vapour degreasing — ICI Ltd

the atmosphere. When not in use, a lid is applied to reduce any loss of vapour through evaporation. When articles are lowered into the hot vapour, the vapour condenses on the cold aluminium and drips down into the bottom of the tank, carrying with it dissolved oil and grease etc., and gradually causing the boiling point of the solvent to rise as the contaminant builds up. Periodically, the contaminated solvent has to be drained off where it can be reclaimed by distillation, usually in a separate vessel, although the operation can sometimes be carried out in the degreaser itself. With some plants, the solvent is continuously purified by distillation in a separate tank.

This method is more effective with components of comparatively thick section, as the degreasing action depends on the amount of solvent condensing on the surface. It readily removes oils and greases from aluminium, but is less effective in removing residues of polishing compounds, particularly if the compounds have been allowed to harden through a delay between polishing and degreasing.

The process is used to remove some protective coatings based on lanolin or other grease. The advantage of the process compared to dipping components in a tank of cold solvent is that the work is only in contact with pure vapour.

2.4.1 Solvents Used

The solvent most frequently employed is suitably inhibited trichloroethylene ("trike"), but other solvents are also used. Apart from the ability of a solvent to dissolve pollutants, other factors have to be considered such as boiling point, toxicity, flammability, and latent heat of vaporisation. It should also have a high vapour density to form a compact layer over the boiling surface and reduce losses from evaporation. Increasingly tetrachloroethylene is being used, particularly in Germany and the United States, which has a higher boiling point, hence its greater ability to degrease thin sheet or to remove high melting point grease. It has the disadvantage of a higher running cost due to the higher cost of the solvent and heating requirements.

As mentioned earlier, pure trichloroethylene is not suitable for use for vapour degreasing aluminium. Under certain conditions it can react with water that may have been induced into the degreaser to produce hydrochloric acid which in turn would react with the aluminium. The reaction can be catalysed when aluminium is present in a finely divided state such as swarf generated during the sawing of extruded lengths. Thus trichloroethylene used for vapour degreasing aluminium always contains inhibitors to prevent the above reaction taking place, and their presence or effectiveness is controlled by a simple titration to determine the presence of acid conditions of the solvent. A number of proprietary grades are available for treating aluminium (e.g. Triklone N*).

2.5 CHEMICAL CLEANING

The use of solvents for cleaning aluminium will only remove surface contaminants which are comparatively loosely held. This may be sufficient for some purposes, but where further processing is carried out, such as painting, anodising or plating, a chemically clean surface is necessary.

In addition to contaminants, there can be present on the surface aluminium or magnesium oxides in varying amounts or in different forms. These oxides have to be removed and replaced by a uniform oxide surface otherwise they can produce variable and unpredicted responses to later surface treatments.

Chemical cleaning is more than simply freeing the surface of contaminants. In most cases it involves dissolving off the surface layers of aluminium, oxides, or heat treatment staining, by a small amount of etching and leaving the thin oxide coating normally present in a uniform condition.

The chosen method of cleaning will depend on three variables:

* ICI Ltd.

(a) alloy or grade of aluminium
(b) surface contaminant
(c) subsequent processing.

In most cases, the aluminium components should be racked on aluminium or titanium jigs to allow the components to be immersed in the cleaning solution, with free access of the solution around all surfaces. The cleaning liquids, usually contained in a metal vessel, are heated to around 50°C. The aluminium should be immersed loosely in bundles, as free circulation of liquid is required to allow easy access and to avoid staining from differential etching. Thorough rinsing after treatment is necessary to remove all residues of cleaner, and provision for draining and drying may also be required.

Examples of different types of cleaners are given below. However, cleaning aluminium, like etching, is a specialised operation, and its success or otherwise will determine the effectiveness of subsequent treatments. Many "failures" in metal-finishing can be attributed to inadequate cleaning and it is a stage that must not be skimped.

That being so, there are a number of proprietary cleaners made by suppliers which, in addition to the basic ingredients, contain a number of useful additives. The suppliers of these cleaners will invariably provide technical assistance on operating and controlling their solutions, often a very important consideration when deciding whether or not one's own cleaner should be used.

2.5.1 Etch Cleaning

Most chemical cleaners for aluminium are alkaline and are based on mixtures of caustic soda, tri-sodium phosphate or sodium carbonate. The caustic soda will saponify grease or lubricants from the aluminium surface, whilst at the same time dissolving a small amount of aluminium, leaving a clean etched surface. The solutions are often inhibited to prevent excessive etching of the aluminium. Caustic soda solutions can be used alone on comparatively pure grades of aluminium, i.e. 1200 aluminium (99% purity). A suitable solution is:

Sodium hydroxide	5%
Temperature	40–60°C
Time	30 seconds–2 minutes.

The cleaning solution is held in a mild steel tank. A disadvantage is that if patches of stubborn contaminant is present, an uneven etched finish may result.

The effects of scale formation can be reduced by adding inhibitors such as gluconates to the caustic solution. These allow the precipitates of aluminium to be produced as a light slurry or suspension rather than formed as hard scale on heating coils and on the tank bottom, reducing the need to chip out when cleaning the tank.

The use of caustic soda as an etch-cleaner on its own can cause problems through differential etching of the aluminium which occurs when patches of contaminant may be extra resistant to the caustic. In these instances it is preferable to use an inhibited alkaline cleaner to remove grease etc. which has little or no etching reaction. It is good practice to precede alkaline etching in caustic soda with treatment in this type of cleaner.

Examples are:

(1)	Sodium carbonate	1.5%
	Sodium bicarbonate	1.0%
	Sodium polyphosphate	0.3%
	Wetting agent	0.2%.

The solution can be contained in a mild steel tank, and is operated at 20–40°C for up to 5 minutes.

(2)	Sodium carbonate	2.5%
	Sodium silicate	1.0%
	Sodium polyphosphate	0.3%
	Wetting agent	0.2%.

This solution is a little more aggressive than the first and is also contained in a mild steel tank and operated at 50–80°C for up to 5 minutes.

2.5.2 De-Smut Solutions

If treatment times in caustic soda etch cleaning solutions are greater than 30 seconds, a smut due to impurities or alloying elements is formed on the surface which in most cases can be removed by a short dip in cold nitric acid (5–50% by volume). The exception is alloys containing silicon where the smut formed is insoluble in nitric acid. In this case, hydrofluoric acid is added to the nitric acid, a suitable formulation being:

Nitric acid	20–50%
Hydrofluoric acid	2–5%.

This solution is more hazardous to health than nitric acid alone and the presence of fluorides can cause blisters on the skin unless accidental splashes are washed off immediately.

The solution should be contained in a steel tank lined with polythene or other acid-resistant coating.

Other de-smut solutions that have been used after alkali etching can be made up from acidic salts such as sodium bisulphate ($NaHSO_4$) and ferric sulphate. The advantage is that dry salts are safer to handle than carboys of strong acids, and toxic fumes are not given off, as can happen with a de-smut tank containing nitric acid. Various proprietary de-smut products are available from chemical supply companies.

2.5.3 Acid Etch Cleaning

Acid cleaning is used for particular applications. Not so effective as alkali cleaners for removing grease or lubricants it finds favour in removing oxide films, which are often thickened during thermal heat treatments of aluminium. It is particularly effective in removing magnesium oxides found on components made from aluminium alloys containing magnesium such as 5251 sheet or 6063 and 6082 extrusions. Magnesium oxide can be formed on the surface of these alloys when annealed or aged, and these oxides are not readily removed in alkali solutions. As a result, uneven etching or cleaning could arise leaving a stained or mottled surface which would not react well to subsequent processing.

The problem does not arise when using acid cleaners, since any magnesium oxide is removed along with aluminium oxide allowing subsequent chemical treatments to take place in a uniform manner. These cleaners are also effective in removing any baked-on lubricants or other carbonaceous deposits.

Typical acid cleaners have the following compositions:

(a)	Sulphuric acid	9.0% (by volume)
	Phosphoric acid	1.0% (by volume)
	Wetting agent	0.2% (by volume)
	Temperature	50–80°C
	Time	1–5 minutes.
(b)	Sulphuric acid	10% (by volume)
	Sodium fluoride	4% (by weight)
	Temperature	Room
	Time	1–5 minutes.

One of the most effective acid cleaning solutions for removing oxides etc. from aluminium is described in Method O of DEF Stan 03-2 (Cleaning and Preparation of Metal Surfaces) which supersedes DTD 915.* It is suitable for most alloys, except those containing more than 6% copper where pitting may result. Parts are first cleaned with an organic solvent to remove grease etc., followed by immersion in the following solution, which produces a lightly etched surface:

Sulphuric acid	150 ml/l
Chromic acid (CrO_3)	50 g/l.

* See Appendix II.

The solution is made up in de-mineralised water in a lead lined tank. Instead of 50 g/l of chromic acid, the equivalent amount (150 g/l) of potassium dichromate may be used. Temperature is maintained between 50° and 65°C, with an immersion time of 20 minutes when used as a pretreatment for anodising, painting or other organic coating. For cleaning prior to adhesive bonding or electroplating, the temperature should be between 60 and 65°C with a treatment time of 30 minutes.

A precaution should be noted. When work is immersed in the cleaning solution, the work or jigs must not be allowed to come in electrical contact with the tank walls, and they must be insulated at the points of support. Failure to do so may result in pitting of the aluminium due to galvanic attack, this being more liable to occur with copper-containing alloys. The method must not be used for assemblies containing other metals, particularly steel or copper, and contaminants such as chlorides and salts of copper and iron must be avoided to prevent pitting.

2.6 CLEANING OF CORRODED ALUMINIUM

Where it is necessary to retreat aluminium components that have become coated with corrosion products, the following method should be used: Components are immersed in the following solution, contained in a stainless steel (EN58J) tank.

Chromic acid	7.5–10 g/l
Phosphoric acid (d 1.75)	5–7.5 ml/l
Temperature	95–98°C

For sheet material, times of up to 30 minutes duration are necessary, whilst castings may require up to one hour.

This solution, described in DEF Stan 03–2, Method U, has the advantage of removing oxides and corrosion products without any etching of the aluminium.

More stubborn corrosion deposits may require the use of the sulphuric and chromic acid solution (Method O of DEF Stan 03–2), described above. This will slightly etch the surface of the aluminium being treated.

Where it is not possible to immerse the aluminium in a heated bath, the metal can be treated *in situ* by application of acid cleaners containing phosphoric acid. A cleaner of this type can be made from a 10% solution of phosphoric acid to which kaolin is added to make a paste (described under CS2625 — "Solution, pretreatment for aluminium"). This is applied by brush, and left for a period to allow the acid to react with any corrosion products, oxides, and to help degrease the surface. After the necessary time to allow the acid to work on the surface, all residues must be removed by copious water washing to remove any residues that might become entrapped in crevices etc.

Proprietary solutions are available, based on phosphoric acid, with added wetting and thickening agents which are used for cleaning down weathered aluminium, such as may be found on neglected buildings, and as cleaning agents for vehicle bodies, rail cars, etc.

These cleaners can be used on bare aluminium, but must not be applied to anodised or painted finishes which would be damaged by the acid. They are often used for cleaning or slightly etching the surface of aluminium that is going to be given a coating of clear lacquer, such as on the outside of transport containers, etc.

2.7 CLEANING ARCHITECTURAL ALUMINIUM

For the more prestigious applications, architectural aluminium is often protected by anodising or by organic coatings. But in some cases, mill-finish

FIGURE 2.3. Before and after. Aluminium piston where burnt-on carbon deposit has been removed by vapour blasting with glass beads — Vacu-Blast Ltd

aluminium is used without any added protection and this often applies to castings.

As time passes the surface may become weathered and if not regularly cleaned the build-up of dirt can allow formation of thick deposits of aluminium oxide which may become unsightly when appearance is important.

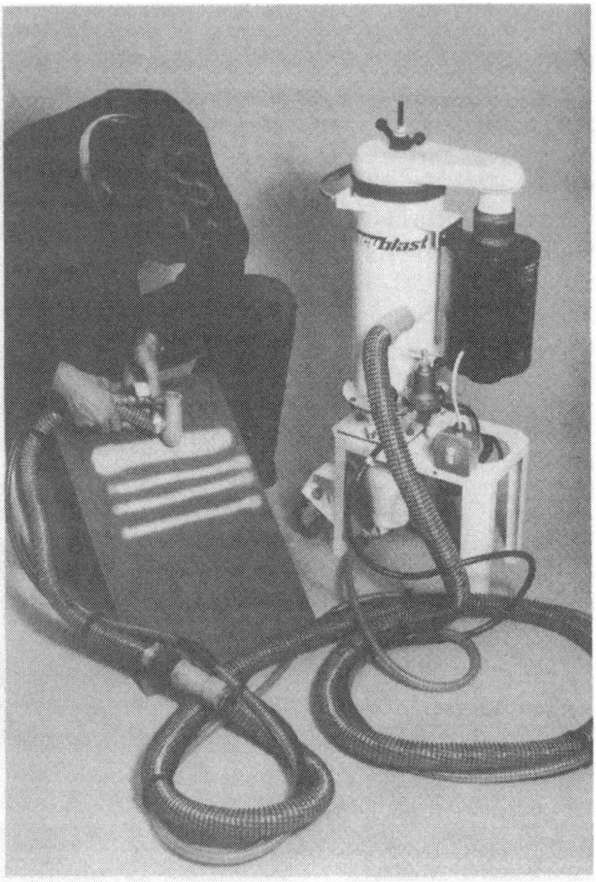

FIGURE 2.4. Hand cleaning of component by abrasive blasting — Vacu-Blast Ltd

Regular cleaning with water, to which a mild detergent such as Tepol is added, will ensure that deposition of dirt will not take place, and will help to maintain the appearance of the aluminium and reduce the rate at which oxides are built up. A soft brush is helpful and this method is also recommended for maintaining the appearance of anodised aluminium. The detergent should be rinsed off with fresh water after the cleaning procedures.

Where the aluminium has been allowed to weather over a number of years without regular cleaning, a considerable amount of oxide and dirt may have built up on the surface. Although not detrimental to the strength of the aluminium components, it may be necessary to restore the original appearance. In this case, the use of phosphoric acid based cleaners can remove the dirt and oxides. Care must be taken to prevent splashing other components, particularly when used on high-rise buildings. After treatment there is the need to remove all residues of cleaner with plenty of water, and the problem of disposing of the washings as phosphates can produce contamination effluent. These cleaners must not be used on anodised or painted aluminium as damage can be caused to the coatings.

Mild abrasives, of the correct formulation although sometimes used on anodised aluminium, are not easy to apply on mill-finish aluminium to produce a uniform finish.

Where aluminium has become badly weathered, or heavily contaminated, it can be renovated or cleaned by one of the abrasive blasting methods described under mechanical finishing (Chapter 1, 1.5), illustrated in Figure 2.3.

High pressure water jets can be effective, but often blasting, sand or alumina can be used, particularly with cast aluminium which tends to be more rigid on account of its greater thickness.

Portable machines are available which provide a jet of aluminium and air to clean and abrade the surface, while around the jet is an external concentric tube which sucks back the debris and spent abrasive. The end of this tube consists of a brush, similar to that on a domestic vacuum cleaner tube, to form a seal to prevent escape of abrasive. (Figure 2.4.)

Chapter 3

Chemical Finishing

3.1 INTRODUCTION

Aluminium, when exposed to normal atmospheric conditions, is highly resistant to change or degradation because of the ever present thin, but protective, layer of aluminium oxide on the surface.

Nevertheless, if the oxide is disturbed, the surface can be changed in a number of ways to alter its appearance, ranging from a matt, diffuse finish (produced by etching) to one of bright mirror-like reflectivity (produced by brightening). These variations can be obtained by treating the aluminium with one of a number of solutions, depending on the application.

When selecting the particular solution to be used, it is important to note the type of aluminium involved. Usually specialised advice is available from the aluminium supplier as well as the producer of the chemicals or proprietary solutions which may be used. Aluminium is produced in a variety of alloys for specific mechanical properties, but a range is also available where particular surface treatments are to be carried out.

Chemical finishing of aluminium is generally chosen wherever possible in preference to mechanical finishing, because the operations are less labour intensive and a higher through-put is possible. It is seldom used on its own without further protection since the treated aluminium surface is easily marked by handling or staining, and it is necessary to follow up such a process by either anodising (which is more usual), or by clear lacquering to protect the surface without altering its appearance. A number of treatments are available to provide for different applications.

3.2 ETCHING

During manufacture, aluminium components may be superficially marked by the various processes such as blanking, forming etc. as well as from the effects of rolling or extruding. In many cases this may not matter, but often it may be desirable to remove or mask any rolling marks on sheet, abrasion or scuff marks or any other surface imperfections. Extrusions may also show directional marking due to the extrusion process.

The effects of these superficial defects can be overcome by etching the components in acid or alkali solutions to produce a frosted or diffuse surface which disguises any scratches or mechanical damage, eliminate glare from an originally bright rolled sheet or give a uniform, but matt appearance.

A typical example of this type of treatment is for aluminium-bodied underground trains where the surface is given a light etch to provide a uniform grey surface and to avoid reflected glare.

Other applications include the etching of aluminium extrusions, such as window frames etc., carried out before anodising. Aluminium beer kegs, made from thick aluminium pressings welded together are often etched on the outside to clean and improve the appearance before despatch to the brewery.

By far the greatest proportion of etching is carried out as a pretreatment before anodising, described in *Anodic Oxidation of Aluminium and its Alloys*, and only a brief description is given here. A freshly etched surface would be easily finger marked or scratched and it is usual therefore to protect it by anodising or by lacquering which preserves the surface without altering its appearance.

3.3 ALKALINE ETCHING

Chemical etching can be carried out in a number of solutions but the simplest and cheapest form are alkaline solutions based on caustic soda. When aluminium is immersed in hot caustic soda the surface is progressively dissolved to form a finish which is microscopically roughened and on which a large number of small pits or depressions are developed. This produces a matt or dull appearance due to scattering of reflected light. The final appearance will depend on the extent of etching and on the size and shape of the pits, as well as contributions made by the aluminium, such as composition and methods of manufacture.

A typical etching solution is:

Caustic soda	5%
Temperature	50–65°C
Time	10–20 minutes.

Various additives are sometimes made to this type of etching solution to change or modify the etching action, and those often used are nitrates, nitrites, fluorides, wetting agents etc.

One of the problems in operating an alkaline etching process is the build up of scale on tank walls and heating coils. This causes the solution to become saturated in dissolved aluminium which is eventually precipitated as alumina in the form of a rock-like deposit. This effect can be reduced by adding chelating agents such as gluconates and heptonates to the etch solution.

The etching process is not as simple as one might at first except, and rather than use a simple caustic soda solution, many operators use one of the specially formulated proprietary products available from one of the chemical supply companies in this field. One of the problems in operating an etch is that even with the special additives the amount of dissolved aluminium builds up in the solution slowing the reaction rate. So-called "Long-Life" etches are used which remain in a workable condition because of the balance obtained between the rate of drag-out from the solution and additions of fresh chemicals added to maintain its activity.

A problem with caustic etching is the amount of hydrogen evolved which produces spray, containing caustic soda, and which makes the working atmosphere unpleasant. Lip extraction or other means of fans and ventilation are necessary to overcome this.

The advantages of alkaline etching are:

1. Economy in use of chemicals and in operation
2. Mild steel tanks suitable
3. No toxic wastes.

The main disadvantages are:

1. Build up of bulky sludge from dissolved aluminium on neutralisation of effluent
2. Exhaust extraction of alkali spray required
3. A process time of 10–20 minutes required.

An undesirable feature of alkaline etching is the formation of a deposit or smut on the etched surface of the aluminium from insoluble constituents of the aluminium and from dissolved constituents in the etch solution. It is necessary to remove this smut by a subsequent short immersion in a 10–20% solution of nitric acid contained in a stainless steel or PVC lined tank.

Nitric acid can give off poisonous fumes of oxides of nitrogen and it is essential to provide good ventilation. A cheaper alternative is to use one of the proprietary solutions, often containing ferric sulphate, which are less hazardous and equally effective in removing smut.

An advantage of the de-smutting solutions is that they neutralise the alkaline film adhering to the surface of the aluminium and remaining after etching, which is not otherwise easy to rinse by simple washing in water.

3.4 ACID ETCHING

As mentioned, because of its greater cost, etching in acid solutions is not so frequently used, but in certain circumstances there are advantages.

Aluminium casting alloys have a high proportion of silicon in their make-up in order to reduce the viscosity of the metal when liquid and to allow the metal to flow easily for the production of casting intricate shapes. The silicon content may be as much as 13%, and being insoluble in caustic soda, a dark grey deposit would be formed on the surface on etching in alkaline solutions. An acid etch containing hydrofluoric acid will dissolve silicon and will not thus produce a smut. A typical acid solution is:

Nitric acid	15% by volume
Hydrofluoric acid	3% by volume
Temperature	25°C
Tank lining	PVC.

This type of solution can be used cold for removing the deposit formed after etching silicon-containing alloys, such as castings, in caustic solutions. For these alloys it may be necessary to repeat the cycle several times as the silicon deposit can inhibit further etching unless periodically removed by the acid dip.

Hydrofluoric acid requires handling with great care and any splashes on the skin must be avoided to prevent blistering. Any effluent must be treated to remove the poisonous fluoride salts.

3.5 ON-SITE ETCHING

The solutions described so far are only suitable for factory use as part of the process cycle where components are immersed in a series of tanks. It may be necessary to treat large fabricated components on-site, such as the underground carriages already mentioned. Similar instances are truck bodies, transport containers and architectural aluminium installed on buildings. Here it may be necessary to provide an even, matt finish to large areas of aluminium and to avoid the effects of fabrication. A number of solutions are available, and although often intended for cleaning or renovating aluminium which for some reason or other has become roughened by weathering, they are also used to alter the surface by removing any contaminants and reducing the reflectivity of the aluminium by leaving a clean, etched appearance.

These solutions usually contain about 10% phosphoric acid, sometimes with additives to thicken the solution to allow them to be brushed on to the surface to be treated where it will remain and react with the aluminium. The solution is allowed to remain on the surface for up to about 10 minutes after which it must be washed off by hosing with copious supplies of water. Containing phosphoric acid, this type of cleaner must be used with care, and protective clothing and eye-shields are necessary. Splashing of other surfaces which might be damaged must be avoided, and due regard must be taken of the contaminated rinse water, as, unless well diluted, effluent problems can arise. The aluminium should then be dried thoroughly before applying an air-drying clear lacquer to protect the surface.

3.6 BRIGHT ETCHING

The etches mentioned previously produce a matt finish with low reflectivity. Certain etches have been developed to produce a bright-etched finish with more lustre and sparkle. The finish produced is still diffuse, as opposed to a specularly reflective or mirror-like finish given by chemical brightening solutions. These have been used for producing reflectors for background lighting, in contrast to the specular reflectors used for spotlights and searchlight reflectors. The finish is often used for cast aluminium domestic hardware e.g. door handles, hinges, light brackets etc. which are otherwise difficult to provide with a decorative effect.

Most of these solutions are proprietary solutions, such as Phosbrite 155 and 156* in the United Kingdom and others of a similar type are also available. They are based on phosphoric acid and sulphuric acid with other additions, and operate at about 80°C. These hot, concentrated acid solutions are vicious in action and considerable care has to be taken to prevent splashing or other contact with the skin. Protective clothing such as gloves and aprons is necessary for the operators safety. Solutions must be contained in lead-lined tanks. No fumes are evolved, but as some spray may be generated during working, exhaust extraction is necessary.

Components are usually jigged on aluminium or titanium racks for treatment which lasts about 2 minutes and which removes about 15 μm thickness of metal from the surface to leave a lustrous and non-directional appearance. After thorough washing, the surface must be protected, usually by anodising but sometimes by lacquering.

3.7 CHEMICAL BRIGHTENING

As indicated the finishes described so far are non-reflective, i.e. although the surfaces may reflect light to a high degree, it is in a diffuse manner and can not be said to be "mirror-like". Generally when one describes a surface as bright, one means it acts as a mirror, and the obvious example is chromium plating with a high specular reflectivity.

* Albright & Wilson Ltd.

Aluminium can be treated to give a similar high specular reflective finish, and when bright bumpers, trim etc. were in vogue for motor cars a large amount was used. The demand for black finishes for trim has currently reduced the amount required but as fashions change reflective finishes could well return. Bright, reflective finishes are also used for lighting reflectors and decorative trim for many household appliances and components.

Mechanically polished aluminium on its own would become oxidised and dull over without further protection, but it does not provide a suitable base for the usual protective methods of lacquering or anodising. This is due to the soft nature of the aluminium, which flows during polishing, entrapping polishing grease and compounds into the surface which interferes with further processing. A chemically clean surface is required, which can be given by chemical brightening, and a range of solutions is available.

These solutions have been sometimes wrongly described as chemical polishing solutions, but they have not sufficient levelling powers to remove anything other than the finest abrasion marks etc. Prior mechanical polishing is necessary for the best finishes, particularly on extrusions which, by their nature, may have directional surface marks.

Chemical brightening solutions have been available from the 1950s and since then have been developed to a high degree. The solutions now used are based on phosphoric acid, and an approximate composition is:

Phosphoric acid	75%
Sulphuric acid	16%
Nitric acid	6%
Copper nitrate	0.5%
Boric acid	0.5%.

Other constituents are also present, and there are a number of suppliers of proprietary solutions, such as Phosbrite* and Alupol.† A number of variants of each are offered for specific applications, such as for automatic anodising lines at many of the larger companies.

Since the solutions consist of a mixture of concentrated acids, they are corrosive and tank construction must be from low carbon molybdenum stabilised stainless steel, such as EN58J (Type 316L) The temperature is maintained at about 100°C, and aluminium components, fixed on racks or jigs of aluminium or titanium, are immersed for about 2 minutes, during which about 10 μm is lost from the surface. Fumes of oxides of nitrogen are evolved requiring effective exhaust extraction which requires scrubbing to neutralise and remove the acid residues.

As the solution is used, the aluminium content increases to an equilibrium value of about 45 g/l as drag-out is compensated for by additions of fresh

* Albright & Wilson Ltd.
† Vernal S.A.

solution. Nitric acid has also to be added to compensate for losses through fuming and the amount has to be maintained closely to the correct level. This will vary with the quantity of dissolved aluminium in the solution, and as this rises the optimum level for nitric acid is reduced.

Although the more commonly available grades and alloys of commercial aluminium can be brightened, they would lose their reflectivity if they were anodised and the anodic coating is increased. However, 99.5% aluminium (1050A) and 99% aluminium (1200) and 99% aluminium containing 0.9% magnesium (5005) as sheet are sometimes brightened where the requirements are not so demanding. Ninety-nine per cent aluminium with 0.4% magnesium and 0.4% silicon (6063) in the extruded form can be brightened.

To make full use of the capability of chemical brightening, a range of high purity alloys based on 99.85% aluminium are used, which can be brightened to produce highly reflective finishes. These are usually alloyed with magnesium to increase the strength. These grades and alloys are produced under particular manufacturing conditions to achieve the best reflectivity.

High Purity Alloys Suitable for Brightening

Form	*Alloy of Grade*	*Designation*
Rolled	99.8%	1080A
	99.8% + 0.4% Mg	
	99.8% + 0.8% Mg	5657
	99.8% + 1.0% Mg + 0.25% Mn	5457
	99.8% + 2.0% Mg	5252
Extruded	99.8% + 0.8% Mg	5657
	99.8% + 0.65% Mg + 0.4% Si	6463

After components are brightened, a deposit of finely divided copper is formed on the surface, which must be dissolved in a bath of 20% nitric acid, or by an acidified solution of ferric sulphate, similar to that used for desmutting after etching.

3.8 ELECTROBRIGHTENING

For electrobrightening, components are fixed to aluminium jigs or racks as for anodising, immersed in the solution and a dc current is applied. Current densities are higher than for anodising and due allowance must be made in jigging to allow for better contacts.

One of the earliest methods was the Brytal process used for treating Super Purity aluminium (99.99%). Extremely bright finishes of high specular reflectivity values around 85% are usual, suitable for specific applications, such as reflectors for searchlights, spotlights and lasers. The introduction of chemical brightening of lower purity alloys based on 99.8% aluminium has restricted its use to only the more critical applications, and the difficulty in obtaining sheet based on 99.99% aluminium is likely to reduce its use further. Modifications in Germany allowed it to be used for aluminium based on 99.9% purity.

The original Brytal process used a solution containing 5% sodium triphosphate and 15% sodium carbonate at about 80°C. Being alkaline, mild steel could be used for the tank, and no corrosive spray or effluents were produced. Processing time was about 20 minutes and a current density of 1.6–2.0 amps/dm^2 was necessary after an initial surge of about twice the steady state. The voltage required ranged from 12–15 with a fixed anode or 18–22 if anode oscillation was used.

Other solutions used were based on hot phosphoric and sulphuric acids or phosphoric and chromic acids. They were useful in that they were more polishing solutions rather than brightening solutions, but they operated at very high current densities.

Electrobrightening is carried out in Germany and the United Kingdom using a solution based on sulphuric and chromic acids, and it is claimed by the users that results by this process are better than by chemical brightening.

In view of its greater flexibility, chemical brightening is now more likely to be used by the general operator whilst electrobrightening will be restricted to the more specialised items.

Chapter 4

Electroplating on Aluminium

4.1 BACKGROUND TO PLATING

Coating metals such as brass, copper or steel with other metals by electroplating is a frequent finishing process, and plating by chromium on to other metals is commonly carried out to improve the appearance, to provide protection or to impart different surface characteristics such as improved hardness etc. It is natural, therefore, to consider electroplating metals on to aluminium for the same reasons, but there are certain differences in treatment which have to be taken into account.

In comparison with plating on to other metals, there are two problems in particular which must be considered. One is the difficulty in applying an electrodeposited metal layer while at the same time obtaining good adhesion between the plated metal and the aluminium. This is due to the oxide layer on the surface of the aluminium which is held tenaciously. The second problem is the possible corrosion hazard present when two dissimilar metals with differing electropotentials are in contact, thereby introducing a galvanic effect to the detriment of the aluminium.

The first problem is overcome by using special pretreatment processes to remove the oxide coating from the aluminium, whilst at the same time replacing it with a thin metallic layer which protects the aluminium from further oxidisation and allows deposition of a metal coating by electroplating.

The second problem is the corrosion hazard brought about by the differences in electropotential between aluminium and most other metals which are less electropositive. If aluminium and, say, copper are in intimate contact, and an electrolyte is present such as water contaminated with a salt or acid, an electric current is generated, a similar reaction which occurs with the so-called dry battery, and in doing so, the aluminium will tend to dissolve, at the expense of the copper. So, when copper is plated directly on to aluminium, any damaged areas resulting in exposure of the underlying aluminium are a potential hazard. Should moisture be present, and especially in the presence of moisture and a contaminant forming an electrolyte, corrosion of the aluminium can occur, resulting in blisters on the coating due to disruption of the bond between the aluminium and the copper or other plated metal and eventual failure of the coating.

It is, therefore, essential that, not only should the plated coatings be complete and free from any defects disrupting the continuity of the coated layer, but that good adhesion between the two metals is achieved.

A feature of all electroplating processes is the throwing power, by which it is meant the ability to coat recessed and other difficult areas with an even and uniform thickness of coating. This is of particular importance where aluminium is concerned, since any lack of protective coating in these areas will prove hazardous. An instance of this is in plating aluminium bumpers where the reverse face of a component cannot easily be coated adequately, requiring a subsequent application of a protective coating such as paint to be applied to these unseen areas after plating.

For these reasons, decorative electroplating of aluminium has been restricted mainly to areas where the corrosion hazards are less severe, such as domestic hollow ware and other non-critical applications. Decorative plating has also to compete with anodising, where special high-purity based alloys can be processed to give a bright abrasion- and corrosion-resistant finish (see *Anodic Oxidisation of Aluminium and its Alloys* by V. F. Henley). However, advances made in the technique of electroplating aluminium, both in the pretreatment stages and in the actual plating processes themselves, have reduced the hazards and made electroplating a more attractive process.

4.2 ADVANTAGES OF ELECTROPLATING ALUMINIUM

An advantage of electroplating over anodising is that ordinary commercial purity aluminium can be treated. The bright finishes that can be achieved with anodising require the use of special high-purity grades of aluminium and special brightening processes. The alloys themselves have to undergo selected manufacturing techniques to achieve homogeneity since the anodised appearance depends on the metallurgical structure. The appearance of plated coatings on the other hand is not affected by the purity of the aluminium, allowing the use of cheaper grades. However, the processing costs of plating are higher than those for brightening and anodising.

Another advantage of plating is that a decorative finish can be achieved with many casting alloys to provide a bright, reflective surface. This finish cannot be obtained by anodising, due to the alloying elements normally present, such as silicon, and to the metal structure developed during casting. Forgings can also be electroplated.

Most plating on aluminium is, however, carried out on wrought products such as those produced from rolled sheet or extrusions.

4.3 EXAMPLES OF ELECTROPLATED ALUMINIUM

Increased confidence in plating on aluminium is evident by the recent erection of a building in Japan where aluminium panels were chrome-plated, rather than being treated by the more conventional anodising methods of finishing aluminium for outdoor use (Figure 4.1). This is the largest application of its kind in the world, situated in an industrial situation not normally regarded as being a mild atmosphere.

Another example of the use of decorative electroplating, under service

FIGURE 4.1. Building clad with nickel/chrome-plated aluminium mirror sheet — Mitsubishi Light Metal Industries and W. Cannings Materials Ltd

conditions which might previously have been regarded as unsuitable, is the use of chromium-plated aluminium bumper bars made from extrusions and pressings in strong alloys normally used for the aircraft industry containing zinc and magnesium (the 7000 series). Plated aluminium in this field has found particular favour in Germany and the USA, where the saving in weight for cars and trucks is considerable.

Earlier examples of the use of decorative electroplating on aluminium are domestic applications such as coffee pots and tableware. The treatment has also been used to decorate aluminium percussion drums.

There are many applications for plating of aluminium used in electrical and engineering fields where it is used to provide a surface with specific properties. Chromium can be applied to provide a hard surface for providing wear resistance in engineering applications.

Hard chromium or hard iron–tin coatings have been applied to the inside walls of cylinders in cast aluminium cylinder blocks and for surfacing of bearings (Figure 4.2). Obviously, for applications such as those where the operating conditions are fairly severe, the requirements of the plated coating and its adhesion to the aluminium must be of high standard.

Electrocoatings of tin or silver are applied to extruded aluminium bus-bars made from 6101A alloy, a magnesium-silicon alloy similar to 6063 but with a high electrical conductivity. The problem with untreated aluminium used for this application is in making joints and connections. Unless precautions are taken, the oxides on the aluminium surface provide increased electrical resistance which would cause joints to overheat and also cause loss of power. A method commonly used is to abrade the mating aluminium surfaces with wire wool and a special grease, the grease preventing re-formation of oxide. The joint has then to be clamped tightly under controlled conditions. These time-consuming operations can be avoided by plating the aluminium with tin or silver, where a conducting and non-oxidising surface is produced with good electrical contact. Either the ends only of the bus-bars are plated, or the complete length can be treated. In the latter case, intermediate contacts can be made along the length of the bar by making tappings at intervals.

Applications of electroplating on to aluminium include the following:

Decorative Applications

- Furniture fittings
- Domestic hollow ware
- Automotive trim
- Cosmetic containers
- Trim for slot machines.

FIGURE 4.2. Aluminium/tin shell bearings for marine engine, plated in tin/lead alloy after Bondal treatment — Glacier Metal Co., Sultzer Bros. AG. and W. Cannings Materials Ltd

Engineering Applications

Printing rolls
Engine cylinder liners and bores
Textile machinery
Moulds for plastics
Bearings
Printing plates

Electrical Applications
- Electrical contacts
- Bus-bars
- Soldering
- Continuous plating of wire

4.4 ALLOYS FOR PLATING

In general, most alloys of aluminium can be plated satisfactorily, but the procedures to be followed in cleaning and pretreatment may have to be varied.

Pure aluminium (1200 series) and wrought alloys containing copper (2000 series) or manganese (3000 series) are the most easily processed. Alloys containing magnesium (5000 and 6000 series) require a modified pretreatment to allow the formation of a suitable electrodeposited coating.

Aluminium castings nearly always contain silicon which would form a smut on the surface of components after alkaline cleaning, not removable by the nitric acid de-smutting solutions normally used. These may require a modified cleaning process. Some casting alloys may also contain magnesium which again require precautions, while LM10, with 10% magnesium, is considered unsuitable.

The responses of the more commonly used alloys to electroplating are to be found in the following table.

Response of some aluminium alloys to electroplating

Alloy	Response	Alloy	Response
Wrought			
1050	V	5083	FX
1200	V	5154	FX
2011	–	5251	FX
2024	V	6063	FX
3103	G	6082	FX
Cast			
LMO	V	LM20	FX
LM2	FX	LM21	F
LM4	G	LM22	G
LM6	FX	LM24	FX
LM9	FX	LM25	FX
LM10	U	LM26	F
LM13	FX	LM27	G
LM16	FX	LM28	F
LM18	FX	LM29	F
		LM30	F

V — Very good response
G — Good response
F — Fair
X — Requires modified pretreatment
U — Unsuitable.

4.5 PROCESSING

4.5.1 Early Pretreatment Methods

As mentioned, the essential requirement for a satisfactory plated coating is the removal of the oxide and replacement with a firmly bonded metallic layer allowing subsequent deposition of electrodeposited layers, or to provide adhesion by mechanical methods.

Various ways were used in the past to achieve a mechanical key. Grit blasting was attempted to provide anchorage, but this was not satisfactory. Chemical etching to provide deep and controlled pits for keying was also tried, but, like grit blasting, the subsequent plated coatings were rough and uneven, and keying was unsatisfactory.

FIGURE 4.3. Aluminium connectors for military purposes, plated with cadmium after Bondal treatment and subsequently olive-drab passivated — A. B. Electronics Ltd and W. Cannings Materials Ltd

A more successful method was to use as a substrate an anodic coating, unsealed and with large pores. This was obtained by using a phosphoric acid anodising process to provide an oxide coating with pores large enough to allow passage of electric current and at the same time form a base layer on to which electroplated metal deposits could be applied.

The process was not easy to operate and although some success was achieved, there were limitations, and the anodising pre-treatment was sensitive to the alloy. This method of pretreatment could not be used for die-castings and only some sand-casting alloys were found suitable.

In spite of the difficulties, attempts were made to plate directly on to aluminium without providing an intermediate layer. One of the methods attempted involved electrodepositing zinc from an alkaline solution, followed by deposits of brass nickel and chromium. If a polished finish was required, buffing had to be done after the nickel deposition, and components had to be heat-treated to provide inter-diffusion between the zinc, brass and nickel layers thereby achieving good adhesion.

There were problems in controlling operations and the process depended on special cleaning systems to be carried out before the plating process. In spite of these, there was an advantage of the electrodeposited zinc that it enabled plating of assembled components containing other metals in addition to aluminium.

4.6 PREPLATING PROCEDURES

In practice, the most commonly used pretreatment processes for plating aluminium depends on chemical deposition of either zinc or tin. There are

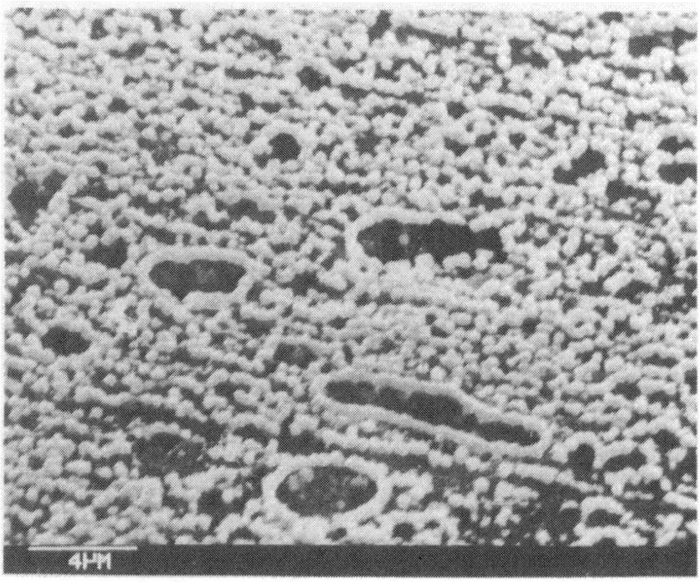

FIGURE 4.4. Electron stereoscan micrograph of aluminium treated by zincate pretreatment process — Dr J. K. Dennis, Aston University

a number of proprietary solutions which have additions to improve the performance of the original zincate solutions and make them more easily controlled. A typical commercial pretreatment used by many platers is Bondal, made by W. Cannings Materials Limited.

A basic zincate pretreatment solution consists of:

Zinc oxide	45 g/l
Sodium hydroxide	425 g/l

Considerable work has been carried out on the variables of time of immersion, temperature and concentrations as well as the effects of various additives. These investigations have brought about improvements to the original basic process which are incorporated in the recommendations and use of the proprietary solutions commercially available. A further advantage of using a proprietary solution rather than one made up oneself once again is the technical assistance available from the supplier. It is also possible to plate some aluminium alloys which hitherto proved difficult to treat.

Figure 4.4 is a scanning electromicrograph of the surface of pure aluminium after 30 seconds immersion by a modified alloy zincate type immersion process. This shows the zinc coating formed on the surface, and the variation in film growth due to structural imperfections such as scratch lines or grain boundaries.

The various practical stages for plating aluminium using a zinc immersion pretreatment are as follows:

4.6.1 Polishing

As much of the polishing to be carried out is done at a later stage, the initial aluminium surface is less critical compared to anodising and the early mechanical finishing operations can be carried out to a lower standard.

4.6.2 Jigging

Components have to be spring-clipped on to metal jigs for providing passage of current and for transporting them through the various treatment stages. Suitable aluminium alloy clips can be used, fixed to aluminium splines. Stainless steel clips are also suitable.

Alternatively, brass or bronze splines can be used with aluminium alloy clips, providing the splines are covered with a protective coating to prevent the possible effects of galvanic corrosion. Sometimes, the clips can be made of bronze, providing the areas on the aluminium adjacent to the contact area is not critical.

After processing and before re-use, deposited coatings on aluminium jigs or contacts can be stripped off in 50% nitric acid, whilst bronze contacts can be given an anodic treatment in 60% sulphuric or 75% phosphoric acid solutions.

4.6.3 Cleaning

Hot alkaline cleaners are suitable, as described in Chapter 2. A suitable solution is:

Sodium carbonate	23 g/l
Tri-sodium phosphate	23 g/l
Temperature	60–80°C
Time	1–3 minutes.

Magnesium-containing alloys are preferably cleaned in an acid etch cleaning solution after first removing any gross contaminants by vapour degreasing or emulsion cleaning. The suppliers of proprietary zincate pretreatment processes are also able to recommend and supply suitable cleaning and etching processes compatible with their zincate process and tailored for use with the alloys being treated.

4.6.4 De-smut

After cleaning, the components are immersed in an acid solution to remove any smut that may have been formed during the cleaning, and to activate the surface of the aluminium. For many alloys to be plated, a 50% solution of nitric acid may be used, 30 seconds usually being sufficient.

For magnesium containing alloys such as 5251 where alkaline cleaning has been used a more effective solution is 15% (vol) sulphuric acid maintained at 90°C, with a treatment time of 2–5 minutes. The process slightly etches the aluminium. When processing cast alloys containing silicon, the shorter time is preferred.

Many alloys are better treated in a nitric-hydrofluoric acid mixture under the following conditions:

Hydrofluoric acid (40%)	10%
Nitric acid (d 1.42)	10%
Temperature	up to 40°C
Time	1 minute.

This is Method W of DEF Stan 03–2. This standard also gives in Method O, an alternative to the fluoride containing acid pickle, being a mixture of chromic and sulphuric acids, described as follows:

Sulphuric acid (d 1.84)	150 ml/l
Chromic acid (CrO_3)	50 g/l
Temperature	60–65°C
Time	30 minutes.

The treatment is said to be only slightly less satisfactory than the acid fluoride pickle, but as the attack on the aluminium is less, a smoother finish is achieved.

Casting alloys with a high silicon content require a more concentrated solution:

Nitric acid (d 1.42)	75%
Hydrofluoric acid (48%)	25%
Temperature	room
Time	3–5 minutes.

The acid solutions containing hydrofluoric acid are more hazardous in use and splashing must be avoided. They should be contained in polythene lined tanks.

4.6.5 Pretreatment

Two pretreatments prior to plating are commonly used:

(a) *Zincate Treatment*. The zincate solution is alkaline, and should be maintained according to the supplier's instructions. A typical solution is contained in a steel or plastic lined steel tank and operated at 30°C. The aluminium is immersed in the solution for up to 2 minutes depending on the alloy. The copper or silicon containing alloys take the longer time. A uniform coating should be produced, this depending not only on the zincate treatment, but on the method of cleaning or acid etching that has preceded it.

The zincate solution is controlled by simple analysis procedures and maintained by one or more additives.

DEF Stan 03–2 gives details of a zinc pretreatment under Method W for aluminium.

1. Dip in 500 ml/l nitric acid (d 1.42)
2. Rinse
3. Immerse for 3 minutes in the following solution at room temperature
   Zinc oxide (ZnO) 100 g/l
   Sodium hydroxide (NaOH) 540 g/l
4. Rinse
5. Repeat stages 1 to 4.

Zincate pretreatment modification. The more easily plated grades or alloys only require a single dip. For the more difficult alloys, such as those containing magnesium, a modified sequence is necessary. This consists of applying only a thin initial zincate coating by immersion for 15–45 seconds, followed by stripping in 50% nitric acid. The zincate immersion stage is repeated for 10–15 seconds to provide a thin but firmly adherent coating.

(b) *Stannate Pretreatment*. An alternative to the zincate pretreatments has also been used for treating aluminium before plating and has found some use in the USA and elsewhere. This is known as the Alstan Process.*

The pretreatment depends on an initial stannate activating solution, followed by a electrodeposition of bronze. It is claimed that this is better suited to overcome any differences or irregularities on the aluminium surface to be treated and will provide better adhesion and greater corrosion resistance.

Prior to the stannate dip, the cleaning and de-smutting operations are recommended by the suppliers. The sequence of operations is as follows:

1. Alkaline clean
2. Acid de-smut
3. Stannate activating solution
4. Electrolytic bronze deposition
5. Conventional plating sequence.

The acid de-smut solution consists of 30% nitric acid and 30 g/l ammonium bi-fluoride for treatment of most alloys. It is claimed that the fluoride is necessary to provide a homogeneous surface.

Castings containing silicon, cast and wrought alloys containing magnesium and manganese containing alloys (3000 series) require an acid de-smut solution containing 3 parts nitric acid and 1 part hydrofluoric acid.

The stannate activating solution is maintained at 27°C in a stainless steel plastic or plastic lined mild steel tank. Components are immersed in the solution for about 30 seconds.

Without rinsing, the components are transferred live to the bronze plating solution consisting of mainly copper, tin, free cyanide and free KOH maintained at 27–30°C. A current density of 1.5–10 amps/dm^2 is applied for 3 minutes.

It is said that the stannate bath does not provide a continuous metallic coating as with the zincate process, but the tin formed by the first solution is incorporated into the bronze layer, which is thus plated directly on to the aluminium.

4.7 ELECTRODEPOSITION

Once a complete zincate or bronze coating has been achieved, electroplated coatings can be deposited. These include hard chromium, decorative chromium, brass, copper, acid tin, silver, cadmium and zinc.

Acid copper can be applied only after first coating the pretreated aluminium with copper from a copper cyanide or pyrophosphate solution. Gold should be applied over an undercoat of nickel.

Chromium plating is carried out for decoration of many indoor applications but for more severe service applications, such as on car-trim, thicker

* M & T Chemicals Co.

coatings are recommended. These thicknesses are given in BS 1224: 1970 (Electroplated Coatings of Nickel and Chromium).

Service Application	*Minimum Local Thickness*	
	Nickel	*Chromium*
Severe outdoor	Duplex 30 μm	regular 0.25 μm
	or	
	Bright 3 μm	microporous 0.8 μm
Moderate, indoor	20 μm	regular 0.25 μm
Mild, indoor	10 μm	regular 0.25 μm

ASTM Def B253–68 "Preparation of and electroplating an aluminium and aluminium alloy by the zincate process" recommends coating thicknesses. Where corrosion is not a hazard, a nickel deposit of 8–13 μm is recommended, otherwise a nickel thickness of 25–50 μm is specified with a chromium deposit of 0.25–0.5 μm. This standard has been adopted as American National Standard 653–1969.

Electroless or chemical plating, whereby metals are deposited on the surfaces of other metals without the use of an electric current, is a process operated by many electroplaters, and is described in Chapter 14 (14.2).

Chapter 5

Chemical Colouring of Aluminium

5.1 INTRODUCTION

Colouring of aluminium by immersion in various solutions has been practised in the past, but the range of colours that can be produced is small, and the colours are not easily reproducible. The abrasion resistance is low, and the colour effects can be removed by rubbing or abrasion. The greater durability and ease of production of the large range of coloured anodic coating described in *Anodic Oxidation of Aluminium and its Alloys* by V. F. Henley (Pergamon Press) has largely superseded colouring aluminium by the methods used in the past which are about to be described.

However, the cost effectiveness of chemical colouring offers an advantage for certain applications such as the identification of small parts, or when the silvery appearance of new aluminium components might be aesthetically unacceptable. It has the advantage of enabling small components to be put in a basket, and lowered into various solutions, without the need to load the individual parts on to a jig, as no electrical contacts are required. Thus it can be an effective method for colouring a large number of components with a minimum of labour, although the alternative methods of anodising or painting are usually preferred.

5.2 CONVERSION COATINGS

These processes are essentially pretreatments carried out on aluminium prior to painting, and are described in more detail in Chapter 6. However, a feature of these processes is that they are normally coloured, and advantage can be taken of this fact to obtain decorative effects, or to reduce glare on architectural features.

Conversion coatings can be divided into three groups, each producing a range of colours dependent on the operating conditions of the solution, but the colour will also depend on the aluminium alloy being treated.

Conversion Coating Type	*Colour*
Oxide Coating — Pylumin Process	grey to grey/black
Chromate-Phosphate — Alocrom 100 (Alodine 100) Bonderite 705	green to blue
Chromate — Alocrom 1200 (Alodine 1200) Bonderite 711	yellow to orange

The Pylumin process, developed by the Pyrene Company, was a development of an older process containing sodium carbonate and sodium chromate. This produces a grey finish, which is partly dependent on the alloy being treated. It has been used on 6082 extrusions intended for military purposes where the dark grey colour is an effective camouflage in the field and to avoid the bright, silvery finish of untreated aluminium.

The chromate-phosphate coatings produced range from green through to green/blue, and it is possible to control the shade to only a very limited extent. Provided colour variation is accepted between different components, it is a useful and cheap method of dulling down the initial bright finish of freshly rolled aluminium building sheet which might cause objection from a Town & County Planning point of view. It also provides a limited degree of protection from surface degradation.

The chromate conversion coatings result in a range of gold to orange colours depending on treatment and the alloy. The colours are again difficult to produce with a consistently controlled finish, and it is difficult to colour different batches with a close match. The colours are not particularly lightfast, and can be removed by the effect of weathering. They have been used on domestic articles such as indoor TV aerials to provide initial sales appeal and where service conditions are not arduous.

Various proprietary processes have been used in the past for colouring but few are now used, anodising being the preferred method of colouring aluminium. Some attempts were made to market so-called "chemical blackening" processes but the practice was never very widespread, as the colour produced was variable and not easy to control.

5.3 THICKENED OXIDE COATINGS

One process recently developed in the United States, "Decoral", and marketed in the United Kingdom by Lea Manufacturing Company, depends on production of a thickened, colourless oxide coating produced by immersion in a weak acid solution for up to 15 minutes. This coating may then be dyed in a similar way to that used for unsealed anodic coatings, and a wide range of colours is possible. It is then necessary to seal the dyed coating by application of a clear lacquer, which adheres firmly to the thickened oxide coating produced by the initial acid dip. The advantage over anodising is the ability to process decoration on small parts in bulk without the need

FIGURE 5.1. Aluminium components coloured by the Decoral process

to provide electrical contact, and avoiding the need to jig individual components. It is a novel way of decorating or identifying small articles, but the colouring produced by this method is not suitable for withstanding direct sunlight which would cause fading. Examples are shown in Figure 5.1.

A possibly cheaper alternative to the coloured Decoral finishes would be the treatment by a colourless conversion coating such as Alocrom 1000, followed by application of a coloured lacquer. This would provide a range of coloured finishes for aluminium which would be suitable for applications in non-critical situations and where resistance to the effects of sunlight was not required. (See Chapter 13.)

Other ways of producing colour on aluminium directly by immersion in various solutions have been used. They have the advantage over dyed processes such as Decoral in that they are less affected by sunlight. Because of inconsistency of colours produced and the dependency on the alloy, they are not now used. Some of the colouring processes that have been operated in the past are as follows:

Black

Potassium permanganate	5–10 g/l
Copper nitrate	20/25 g/l
Nitric acid	2–4 ml/l

The parts to be treated, after cleaning, are lowered into the above solution at 80°C for 20–30 minutes. A black colour is developed, but the surface of the aluminium may have to be brushed with a stiff brush soaked in the solution at intervals to ensure an adequate colour being developed.

An alternative solution is:

Ammonium molybdate	10–20 g/l
Ammonium chloride	15 g/l
Temperature	100°C

A deep black can develop in 1–2 minutes.

A variation on the potassium permanganate–copper nitrate solution will produce a dull brown colour in 10–15 minutes, again at 80°C.

Brown

Potassium permanganate	5–10 g/l
Copper nitrate	5 g/l
Nitric acid	2.4 ml/l

Blue-Bronze

Potassium ferricyanide	5 g/l
Ferrous chloride	5 g/l

Iron Grey

Potassium sulphide	25 g/l
Vanadium sulphate	1 g/l
Temperature	80–90°C

Reddish Brown

Potassium permanganate	320 g/l
Cadmium nitrate	100 g/l
Ferrous chloride	20 g/l
Nitric acid	20 g/l

The colours produced by the above means are affected not only by the particular alloy being treated, but also can depend on the heat treatment if applicable. In particular, components made from most of the casting alloys contain silicon which tends to produce dark shades when attempts are made to colour them in this way. They also can have a powdery appearance. Metal flow lines formed during solidification, or porosity, tend to reduce uniformity.

Chapter 6

Conversion Coatings as Pretreatments for Painting and Protection

6.1 INTRODUCTION

The normal oxide coating present on the surface of aluminium is only very thin, but unlike oxides formed on steel, it provides some protection to aluminium. The naturally formed aluminium oxide does not provide a key for promoting adhesion with applied coatings of paint or adhesives, but fortunately, methods have been developed either to produce an oxide coating which is amenable to painting, or to provide a thin layer of mixed aluminium salts on the surface which will then form a good key.

6.1.1 Oxide Coatings

The early processes consisted of treating the aluminium in alkaline solutions containing dichromates to thicken the oxide coating on the surface, but processing was slow. These coatings were intended to provide added corrosion protection although not all of them were of benefit for subsequent painting, and they have been mostly replaced by various proprietary solutions which have proved extremely effective in providing good paint adhesion.

One of these early systems was known as the *Modified Bauer Vogel Process (MBV)*.

Sodium carbonate (anhydrous)	2–5%
Sodium chromate (anhydrous)	0.5–2.5%
Temperature	90–100°C
Time	3–5 minutes

The solution etches the aluminium and although precleaning is not necessary for some applications heavily contaminated surfaces should first undergo a cleaning operation. Alloys containing magnesium should be given a dip in 20% nitric acid before treatment.

The colour produced is dependent both on operating conditions and the aluminium alloy. Pure aluminium will provide a light grey colour, whilst others will be coloured dark grey or black. The colours are not easy to control nor are easily reproducible.

The main advantage of MBV coatings is the added corrosion resistance that the thickened oxide coatings give. Improved corrosion resistance is provided by sealing the work in a 3–5% solution of sodium silicate at 90°C for 15 minutes. MBV coatings required for subsequent painting should not be sealed.

6.1.2 EW Process

A further modification of the MBV process depended on the addition of sodium silicate.

Sodium carbonate	47 g/l
Sodium chromate	14 g/l
Dry sodium silicate	0.06–1 g/l
Temperature	90–100°C
Time	8–10 minutes

The range of colours produced are similar to those generated by the MBV process.

Experience has shown that the EW process provides better corrosion resistance with a number of substances compared with the MBV process. The former is also more suited for treating strong alloys containing copper.

The MBV process, being more porous, provides coatings with better adhesion for paint, and they can be dyed to provide a limited number of dark colours.

6.1.3 Pylumin Process

This was developed by Pyrene Company Limited as a further modification to the MBV process made by adding to the solution a carbonate of a heavy metal. The advantages over the original MBV process are that:

(1) The solution has a longer working life, and, unlike the MBV process, can be regenerated by addition of further chemicals.
(2) Strong alloys, i.e. those containing copper, can be treated.

The solution is operated at 70°C, with a treatment time of about 4 minutes.

The colours produced by the process are dark grey, and although the coating is intended as a paint base, it is used for colouring aluminium used for military purposes.

6.2 COMPLEX SALT CONVERSION PROCESSES

The original and most well-known process for producing chromate-phosphate coatings was developed in the United States by the American Chemical Paint Company (Amchem), under the name Alodine. It was developed further and marketed in the United Kingdom by ICI Ltd. as Alocrom. The patent rights of many of these solutions have now expired, and there are now a number of similar proprietary systems on the market, such as the various solutions produced by Pyrene Chemical Services Limited under the generic name of Bonderite.

6.2.1 Chromate-Phosphate Process

Alocrom (Alodine) 100 was the original proprietary solution, although other variations were developed later such as Bonderite 705.

The solutions are mixtures of acids, mainly containing chromates and phosphates with some fluorides. The coatings, produced on the aluminium are complex and are more akin to oxides rather than crystalline salts. The colour of the coating is a blue-green on pure aluminium and on the 3000 manganese containing series, and an olive green colour on the copper containing alloys. The desirable coating thickness is in the range of 7–20 mg/dm^2, the thinner coatings being more suitable for paint pretreatment.

6.2.2 Operation of Process

When operated as a dip process, the solution is maintained in a stainless steel bath, with facilities for heating to a temperature not greater than 50°C. Exhaust ducting to the bath should be fitted, as some spray is given off during operation. The solution can be operated cold, with a treatment time of 5 minutes, or 1½ minutes at 50°C. The preferred temperature range is 30–40°C for 2–3 minutes when treated by immersion. Work can be loaded in bundles provided good circulation of liquid is maintained around the components, but preferably the work should be racked on stainless steel or aluminium jigs if subsequent painting is to be carried out.

On removal from the process tank, the work is rinsed for a short time (10–15 seconds) followed by a rinse in a weak solution of chromic acid (0.05%) at 40–50°C in de-ionised water. Finally, the work is rinsed in demineralised water and dried, preferably at a temperature of 40–65°C, but no higher.

When applied by spraying, more concentrated solutions can be used which reduces treatment time to around 20 seconds, which is of considerable advantage in automatic painting lines, either for extrusions or in coil-coating lines. A variation of Alocrom 100 used for application by spray allows it to be dried at up to 140°C, without detriment to the adhesion of subsequent paint coatings.

The make-up of the chromate-phosphate bath consists of chromic and phosphoric and hydrofluoric acids, which are supplied as powders and liquids. The full operating sequence that should be followed out is given as follows:

1. De-oxidise in 20% nitric acid and rinse (This operation is preferred with magnesium containing alloys such as 6063, otherwise it may be omitted.)
2. Etch clean in alkali solution
3. Rinse
4. De-smut in nitric acid
5. Rinse
6. Conversion coating immersion
7. Static water rinse
8. Continuous water rinse
9. Static rinse in 0.01% chromic acid
10. De-mineralise water rinse
11. Dry at up to 65°C.

Some solutions can be applied by spray for automatic conveyorised lines, using spray boxes rather than immersion tanks for faster and more efficient treatments.

6.2.3 Control of Solutions

Control of the conversion coating solution is made by titration with 0.5N sodium hydroxide, the bath is very dependent on its pH, the chromium phosphate and its fluoride content. The titration will determine the additions required, except for the hydrofluoric acid. The fluoride content is not easy to determine, but additions of hydrofluoric acid can be made according to the weight of the conversion coating produced on the aluminium.

A practical method of assessing the activity of the fluoride is to immerse a glass slide in the treatment bath, and observe the degree of etching achieved in a certain time.

Coating weight determination can be made by immersing a weighed panel of conversion coated aluminium of known area in concentrated nitric acid until the coating is removed, i.e. between 15 seconds and 3 minutes. On drying and re-weighing the coating weight can be determined, which should be between 0.08–0.13 g/m^2.

The above method of coating weight determination is only suitable for a freshly deposited coating. Aged conversion coatings require to be removed by immersion in a molten salt bath, such as is used for annealing aluminium. One method, ASTM B-449-67, requires a mixture of salts of the following composition:

Sodium nitrate	99%
Sodium hydroxide	1%

If this method is used, the salts should be fused in a steel crucible until molten which will be about 500°C. Great care must be taken in avoiding moisture, and the aluminium test piece must be thoroughly dry.

After immersion in the molten salts, the sample is removed, cooled and washed in hot water to remove the salts before drying and re-weighing.

As an estimate of the activity of a bath, it is also possible to determine the weight gained by a sample when treated by the solution.

6.2.4 Chromate Process

This process produces coatings, mainly of chromates but with some fluorides. The coatings vary from pale gold to orange-brown depending on alloy and thickness. The very thinnest coatings can be colourless, allowing the application of clear lacquers to be applied. The coatings in general are thinner than the chromate-phosphate processes, consequently the colour produced may be irridescent. Although the corrosion resistance of the chromate coating is intrinsically higher than the chromate-phosphate processes, the latter are preferred as a base for paint due to the increased thickness. Examples of the chromate process are Alocrom 1200 and Bonderite 711.

An advantage of the chromate processes is that the chromium is present in the tri-valent state rather than the hexavalent state as with the chromate-phosphate processes. The tri-valent chromium is non-toxic, and on this account it can be used as a pretreatment for aluminium that is to be lacquered and which is to be used in the food or drink industry.

On account of the corrosion protection given to aluminium, it sometimes can provide an alternative to anodising, particularly with fabrications containing other metals which would otherwise require masking during anodising.

Another advantage is that the coating, unlike anodising, does not appear to affect fatigue properties.

The US Military Specification MIL-S-5002A allows the use of chromate coatings as an alternative to anodising for protection of aircraft parts other than on those on the exterior of seaplanes.

Chromate coatings are used in many industries for providing protection to aluminium as well as providing a base for painting.

Chromate coatings, as with chromate-phosphate coatings, allow the aluminium to undergo spot-welding although contact resistance is slightly higher than for untreated aluminium. Different proprietary solutions may produce coatings with different electrical resistances, and this is of importance in the car industry.

6.2.5 Operation

Tanks and equipment should be made from stainless steel or mild steel protected by PVC. The proprietary solutions containing chromates and fluorides are added to water, and the solution heated to the operating temperature. Solutions can operate between 15 and 65°C, depending on the make-up and the time of immersion required.

The solution can also be applied by spray, and when used on strip coating lines, the excess coating can be removed from the metal by squeegee rollers and dried, without rinsing.

After immersion, the aluminium should be rinsed, followed by a dip in a weak solution of chromic acid and dried at a temperature not exceeding 65°C.

Control is made by titration which decides on the amount of additions required.

A feature of the chromate processes is that, unlike the operation of the chromate phosphate solution, sludge is not generated during use. This avoids the necessity of filtration of the bath.

6.2.6 Effluent Control

Both processes require effluent control to avoid contamination of effluents. With the chromate-phosphate solution, the hexavalent chromium containing rinse waters have to undergo a reducing treatment with sulphur dioxide to convert the chromium to the trivalent state, followed by additions of lime to control the pH and precipitate the heavy metals.

One plant uses a de-ionising system to separate out and recover the chromates from the effluent.

6.3 "NON-RINSE" PROCESS

A modern process has been developed for use particularly on continuous coil-coating paint lines.

The solutions used are mainly mixtures of chromates and finely divided silica. They are applied directly on to the cleaned aluminium coil, either by dipping but mostly by reverse roller coating. The coating is then dried without further rinsing, leaving a fine deposit of silica and chromates on the aluminium surface which provides the necessary key to subsequently applied paints.

An advantage is that the solution is always fresh and unaffected by dissolved aluminium when applied, and no rinse waters contaminated by chromates are produced.

One solution being used in this country is Accomet C. developed by Albright & Wilson and produced by Pyrene.

6.4 PHOSPHATE PROCESS

Fabricated aluminium assemblies may include steel components, typical examples being found in the car industry where aluminium panels may be joined to steel frames before painting. It is necessary, therefore, to use a conversion coating which is compatible with both metals. Phosphate coatings have been well tried for treating steel, and they have been adapted for use with aluminium as well. These are solutions based on phosphoric acid and zinc salts, with other additions, such as fluorides and oxidising agents. Typical solutions of this type are produced by various manufacturers such as Pyrene chemical services who produce a range of zinc phosphate solutions in their Bonderite range.

The solutions can be used by immersion, but are commonly applied by spray on an automated line. During use, the fluoride content is used up at a faster rate than the other constituents, and has to be replaced by the addition of hydrofluoric acid. The rate at which it is used depends on the proportion of steel being treated.

This type of pretreatment is of particular interest to the car industry where the use of aluminium is increasing.

6.5 OTHER PRETREATMENTS

The use of the aforementioned processes involve chemicals which result in effluent problems requiring special treatment. The standards of control for effluent which can be discharged to rivers or sewers are becoming increasingly high, and the purification methods are a costly burden to the producer.

Various Bonderite processes are available which are chromate-free and, therefore, not so troublesome with regard to effluent control; unfortunately, the adhesion they provide to paint is not so effective.

Anodising is used as a pretreatment prior to organic coating. It has been successful for many years on a strip line producing lacquered aluminium for fish cans in Norway. The original paint line installed a number of years ago in

South Wales by Alcan also made use of an anodic coating as a pretreatment before painting. The modern Alcan lacquer coating line at Goettingen uses a patented ac anodising process with a phosphoric acid electrolyte for pre-treatment.

At present, anodising is used by one manufacturer for treating the internal surfaces of aluminium beer kegs, which is used to provide a key for lacquering. The requirements of the coating are that it should be thin (about ¼–½ μm) and it should not be sealed. The aluminium should, of course, be well cleaned before anodising and the coating should be complete over the whole surface.

Sealed anodic coatings of 2 μm and more do not provide a good base for paint, and in the past some difficulties were found when attempting to apply paint to part of the surface of decorative anodised aluminium car trim.

6.6 DEVELOPMENTS

Developments in the United States have suggested the use of complex metal salts of zirconium or titanium with oxidising agents to overcome the need for chromium containing pretreatments. Some production lines are already using them, the coatings produced being colourless and provide reasonable adhesion and corrosion protection. They have the advantage of operating at room temperature. At present, they are not as effective as the chromium containing solutions.

The determination of the coating weight of conversion coatings is of particular concern on automatic coil coating lines where faults in production can necessitate costly rectification, if not detected as soon as possible. An instrumental method now being used depends on the fluorescent radiation characteristic of chromium or other specified elements after subjecting the surface to an X-ray source. The X-ray equipment is portable and safe in use.

Chapter 7

Protective and Decorative Air-drying Coatings

7.1 INTRODUCTION

Painting of aluminium has become of greater interest in many areas largely because of the greater application of this metal. Although in recent years there has been a great increase in the use of the more durable stoved finishes, there are a great many occasions in which an air-drying paint is more suitable, particularly where large fabricated constructions require painting on site. Applications include containers and bodies for road and rail transport, buildings, road work construction and large fabricated components.

Finishing by air-drying painting processes is more flexible, does not involve large capital expenditure for plant and equipment, and the coatings themselves can be cheaper than those requiring thermal curing or stoving. For many articles, the expense of applying a durable stoved finish may not be justified, and a low cost, air-dried coating may be adequate.

A feature when applying paint to aluminium is that the naturally formed oxide coating, normally of benefit with regard to the protection it imparts, does not provide a good key to applied organic coatings. The quality of a paint system is very much dependent on the adhesion of the coating to the substrate, but the natural oxide interferes with the bonding that might otherwise occur. Aluminium oxide is comparatively inert, and it can exist in different forms and thicknesses, which can be affected by the type of thermal treatment given to the aluminium before painting. Its presence may give rise to unpredictable results in the adhesion to subsequent paint coatings. Certain alloys contain magnesium, such as 6063 extrusions or 5251 sheet, the magnesium oxide can present problems which may require additional steps in the normal pretreatment procedures to remove it.

It is essential, therefore, to remove these oxides from the aluminium and either re-form them in a uniform manner, or preferably replace them with another surface layer which will provide the necessary ability to bond with paint.

It follows, therefore, when applying paint or other organic coatings to aluminium, that correct procedures are carried out before painting, and this includes attention to methods of cleaning and pretreatment. It is these features which are important when dealing with aluminium, and once these operations have been successfully carried out, subsequent paint application is similar to that for other metals.

Painting is generally carried out for the normal reasons of providing a coloured surface to make the aluminium aesthetically acceptable, rather than for protecting the metal. Nevertheless, because of the basic durability of the aluminium, a paint coating will usually last longer than when applied to a less durable base, such as steel. Paint is also applied to aluminium to protect it from a corrosive environment.

7.2 APPLICATION OF AIR-DRYING PAINTS

With the exception of lead-based paints, which are seldom used now because of health hazards, most paints in common use are suitable for use with aluminium.

Painting aluminium on-site and in outside environments requires the usual precautions, i.e. removal of dust or grease, a suitable environment, etc. and the normal methods such as by brush, roller, or spray can be practised. The following procedures should be adopted to obtain the best results where maximum life between repainting is required, especially on the exterior surfaces of buildings and other structures.

7.2.1 Cleaning

Methods of cleaning are described in Chapter 1, but the usual method for on-site cleaning of aluminium is by wiping down with clean rags moistened with a solvent compatible with the coatings to be applied subsequently, i.e. white spirit, when applying household-type paints, or thinners when cellulose paints are to be applied.

The phosphoric acid based cleaners described in Chapter 1 are effective in pretreating aluminium as they have a slight etching effect, and are also useful in removing from the surface magnesium oxides from magnesium-containing alloys. These cleaners must be rinsed off thoroughly after use and the aluminium must be dried efficiently before application of subsequent paint coatings.

Conversion coatings, as described in Chapter 6, are applied mainly by immersion or spraying which is suitable for use only in factory conditions. Some can be applied by brushing for on-site treatment but apart from occasional use when repairing previously damaged paint coatings, they are not often used in this manner.

It is possible, however, to obtain aluminium, usually in sheet form, which has already been pretreated by a conversion coating process. This is intended for paint application without further pretreatment other than simple degreasing.

Conversion coated sheet that is required for painting must be entirely grease-free, and it may be necessary, either by vapour-degreasing or wiping down with solvent, to remove any contaminants that may have arisen during fabrication such as guillotining or cutting operations. Providing this is carried out satisfactorily, extremely good bonding can be obtained between air-drying paint and aluminium.

7.2.2 Etch-Primers

Where one cannot use a conversion coating process on aluminium, the most practical means of pretreating aluminium is to apply a coating of an etch-primer, sometimes known as a wash-primer. This is often based on a 2-pack system, requiring mixing just before use. The main solution contains chromates, to which is added the correct amount of activator, containing phosphoric acid. The two solutions are thoroughly mixed shortly before application, although it has the disadvantage of a limited pot life after mixing. A thin coating is applied, usually by brush or roller, forming a light green or yellow finish. It is only necessary to ensure all the surface has been evenly covered, and no attempt should be made to try and produce a thick coating, or worry about uniformity. The etch-primer is allowed to react with the aluminium, and to dry. This can take up to 12 hours, but after this period the coating to follow must be applied as soon as possible. Too long a delay should be avoided, as the etch-prime coating, being hygroscopic, may tend to absorb moisture from the atmosphere and affect subsequent paint adhesion.

It is good practice to allow the etch-prime coat to dry for 12 hours, and to apply the next coat within the next 4 hours. One-pack etch-primers are available, but are not so effective.

7.2.3 Chromate Metal Primer

The second coating to be applied after the etch-primer or the conversion coating is one which again contains chromates, usually zinc chromate. The chromates are inhibitors and being slightly water soluble, help to form a protective barrier should any water diffuse through the paint top coats, and assist in the prevention of any reaction with the aluminium below the paint layer.

The primer is applied by brush, roller or spray to a thickness of about 15 μm, and will adhere firmly to the etch-primed surface of the aluminium, giving a yellow appearance.

The coating should again be allowed to dry, followed by the final paint coats.

7.2.4 Finish Coats

The aluminium may be finally coated with a conventional good quality air-drying one or two-pack coating. To provide a uniformly coloured finish, it may be necessary to use first an undercoat which will provide a dense uniform base on to which one can then apply the final finish. A gloss finish is more durable than a matt finish.

In summary, the optimum sequence for painting aluminium that is to be used on architectural buildings, road-barriers, etc., where a long life is required without having to repaint, is as follows:

1. Clean and degrease
2. 1 coat etch-primer, 5 μm. Dry for 12 hours
3. 1 coat chromate metal primer, 12–16 hours after etch priming
4. 1 coat undercoat
5. 1 or 2 coats finish coat.

7.3 PAINT TYPES

Most air-drying finish coats are based on alkyds, acrylics or polyurethane resins, although other coating types are used for particular applications, e.g. chlorinated rubber paint for acid containing environments. As has been mentioned, lead-based paints must not be used over aluminium as the lead compounds can react with aluminium and result in corrosion.

There are many types of paint which are used in special situations for specific reasons, and some of the applications are described later.

7.4 WEATHERED ALUMINIUM

For painting aluminium that has been exposed and may have become weathered, it is of benefit to vapour or shot blast the surface before cleaning to remove any oxidation products. In the case of castings, it may be necessary also to remove any surface inclusions or debris. Phosphoric acid-based cleaners, described in Chapter 1, can also remove oxidation products and provide a sound base for painting.

7.5 INDOOR APPLICATIONS

Some components may not require the full painting sequence, such as when they are to be used indoors, or when they only require painting for identification, and the complete process is not justified on cost grounds. It would then be permissible to omit either stages 2 or 3, and only to apply one or two finish coats on either an etch-primed or a conversion coated surface.

7.6 ALUMINIUM CAR BODY SHEET

The original finish on car bodies, as applied by the factory, is usually a stoved acrylic paint. Where attempts are made to repair cars using aluminium sheet, and for the painting of aluminium lorry or van bodies, it is more usual to apply rapid drying cellulose paints, usually by spray but sometimes by brush. These paints are also used for coating aluminium, but again an initial treatment with an etch-primer must be used followed by a chromate metal primer before applying the cellulose based top coats. It is necessary to ensure that the primers are compatible with the cellulose finishes that will be used to provide the top coats.

7.7 MARINE APPLICATIONS

Aluminium is frequently used for shipping applications both above and below the waterline, where the magnesium-containing alloys such as 5251 and 5154 are very durable. Providing the recommended pretreatment stages of cleaning, etch-priming and chromate metal priming are followed, painting can be successfully carried out. As on other materials, paint will not last so long as in inland situations before repainting is necessary, but the durability of aluminium will ensure a longer life to the paint.

A satisfactory finish paint for small boats in particular is a polyurethane finish, the most suitable being a 2-pack type which can be applied by spray or by brush.

This provides a hard coat with good resistance to abrasion, the sequence being:

1 coat etch-primer
1 coat polyurethane primer
2 coats polyurethane finish.

The use of anti-fouling paints for small aluminium hulled craft should be avoided, as they usually contain copper which could form a corrosion hazard. Some of the newer organo-tin containing anti-fouling paints are also suspect on this account.

7.8 AVIATION

Painting of aircraft is a specialised field, and many of the materials and processes are covered by Ministry of Defence specifications.

Pretreatment of the aluminium is similar to that for other applications, such as use of a conversion coating or etch-primer. Anodising is sometimes used as a pretreatment, usually in a chromic acid electrolyte, which is an effective method of ensuring good adhesion for paint coatings. Ministry specifications call for the chromic acid process rather than by the more usual sulphuric acid process, since any chromic acid electrolyte entrapped in metal pores or crevices in assemblies will not be a corrosion hazard. The anodic coatings, in this case, should not be sealed after anodising as is usually done for maximum protection, since adhesion of paint coatings will be reduced.

A range of finish coats can be applied, including epoxies, polyurethanes and acrylics.

Apart from the normal hazards of mechanical damage, the paints used have to withstand severe service conditions such as the spillage of various fuels, lubricating oils or hydraulic fluids which may come into contact with the painted surfaces.

The aluminium alloys used in aircraft construction are primarily selected for their mechanical properties, consequently most of them contain either copper or magnesium and zinc. The copper containing alloys in particular must be protected from corrosion, which must be provided by a paint which will not degrade and continue to give protection over a considerable period. The paints also must have good resistance to abrasion and mechanical damage that might arise during handling of the aircraft on the ground.

7.9 PAINTING FOR PROTECTION

Aluminium may be used in situations where it is necessary to protect it under corrosive conditions, either temporarily or permanently. It is then necessary to provide protection by providing a barrier layer of protective paint between the aluminium and the reactive agents. Hazardous conditions could arise in chemical plants or in electroplating shops where the atmosphere is humid and acidic or alkali gases or spray may be present. Problems occur when moisture condenses from warm atmospheres which is readily contaminated by the pollutants causing a reaction with the aluminium. One case in point is concerned with public swimming pools, which are large heated areas containing an atmosphere with a high degree of humidity and chlorine gas that is used for purifying the water. Condensation readily occurs on cold surfaces such as on window frames that may not have been adequately thermally insulated from the outside and on all metal work generally when the temperature is allowed to fall if heating is not maintained throughout the night.

Other aggressive situations can arise in factories, etc., where acid or alkali fumes, or various chemicals, may be present. In these circumstances, it is necessary to paint the surface of the aluminium with a suitably chemically resistant coating. The effectiveness of this type of protection depends on the resistance of the coating which should have the following properties:

(a) good durability and chemical resistance
(b) adequate thickness
(c) low permeability to moisture
(d) good adhesion with the aluminium.

The coating must be inert to the corrosive media and should not be reacted by them. The coating must be stable under the conditions of service, i.e. at the temperature of operation.

The thickness of the coating must be sufficient to provide the necessary barrier, and this should be coupled with low permeability or porosity of the paint. This latter point can be hard to achieve as all paints are permeable to a small extent, but the effect can be minimised by the coating having adequate thickness, if necessary by the application of extra coats.

The adhesion of the coating to the aluminium is important as it must remain firmly attached to the metal to avoid or reduce blistering at any areas of pinholing or through accidental mechanical damage arising in service. To achieve the desired thickness, better adhesion is obtained by applying several thin coats rather than a fewer number of thick coats.

The physical properties of the coating are also important, and as well as the above properties, the coating should be sufficiently hard and strong to withstand normal wear and abrasion that might otherwise cause mechanical damage. Methods of determining the properties of coatings are described in Chapter 12.

A number of specialised coatings are available to provide protection to aluminium which has to be exposed under conditions for which it would not otherwise be suitable. There are also a number of firms who provide a service in the protection of metals used in hazardous areas, such as, for instance, the aforementioned plating shops and chemical works.

Some of the paints that are used include the following types:

7.9.1 Bitumen Paint

Bitumen paint is cheap, being a derivative of the coal and oil industries, and is capable of being dissolved in solvents which, when applied to a surface, will dry to form a chemically resistant film. It is applied directly to aluminium as a thick coating and is particularly useful for providing protection to the metal against wet alkaline cement or mortar when used on buildings.

Aluminium lamp posts are often embedded in concrete, and that part of the metal in contact with the concrete should be given protection whilst the concrete is setting, and until the chemicals and salts corrosive to aluminium contained in the cement are eventually leached out. A coating of bitumen applied to the surface of the posts below and just above the concrete level has proved a low cost and effective method. It is also used in coating the foot of stanchions for hand-railings where the aluminium is to be bolted on to concrete.

7.9.2 Epoxy-Bitumen Paint

A slightly more resistant coating is the result of combining bitumen with epoxy resins to provide a protective system for use in acid conditions often found in chemical plants. The coatings are available in a range of colours, as well as the more familiar black. The coatings provide good protection but should not be used under service temperatures of more than 100°C.

7.9.3 Chlorinated Rubber Paint

This is a paint coating based on chlorinated rubber. The paint, that should be applied as a thick coating, has very good chemical resistance and can be used for protecting aluminium in aggressive environments. It can be used under humid conditions and provides protection against attack from dilute acid or acid fumes. It can also provide protection to aluminium that is to be continuously immersed in fresh or sea water, but it should not be used for prolonged resistance to strong alkalis or certain organic solvents.

For maximum protection, a complete sequence would be as follows:

1 coat two-pack etch-primer
1 coat chlorinated rubber metal primer
1 or 2 coats chlorinated rubber thick coating
1 or 2 coats chlorinated rubber finish coat.

The total dry film thickness will depend on the method of application, a minimum of 300 μm by brush or conventional spray, and 350 μm will be achieved by airless spray.

7.9.4 Epoxy Paint Finishes

These are two-pack systems which are resistant to immersion in many organic solvents and alkalis and can withstand splashing from many of the stronger acids. The protection against acid attack or prolonged water immersion is not as good as that given by chlorinated rubber.

The recommended paint sequence for coating aluminium is:

1 coat two-pack etch-primer
1 coat epoxy metal primer
2 or 3 coats two-pack epoxy finish.

Brushing is preferred for application of the primer to achieve maximum adhesion, although spraying can also be used for the other coatings.

Re-coating between coats of the epoxy finish can be carried out after allowing to dry for 12 hours, but maximum curing will depend on the temperature, and can take up to 7 days at 13°C, after which maximum chemical resistance will be achieved.

It is especially important to carry out correct procedures for protecting against corrosion when aluminium is used in environments which might be particularly hazardous. Painting could be the answer, but there are other alternatives such as plastic coating described in Chapter 10.

Chapter 8

Application of Thermosetting (Stoving) Liquid Paints

The coatings covered by this chapter are those which depend on the application of heat, not only to dry the paint, but to bring about a chemical reaction between the components of the resin to promote bonding or interlinking, and to provide the adhesion between the paint and the aluminium. Air-drying paints, on the other hand, are those that rely on the evaporation of solvents in the paint to harden the coating, a process which may also be assisted by oxidation of some of the constituents. Also included in the category of air-drying paints are those which are known as 2-pack systems, where two components are added together before application, and the paint hardens as a result of a chemical reaction taking place at ambient temperatures.

Coatings in this category are those which are applied during manufacture possessing enhanced properties such as durability or hardness providing a long service life.

There are many situations where the coatings used for aluminium must have better properties that can be provided by air-drying paints. These include architectural finishes on those parts of the building which cannot easily be reached for maintenance or re-painting, hence the longest life finish that can be economically provided. Other instances are harder and more abrasion resistant coatings used to finish domestic kitchenware etc., where both decoration and freedom from damage is wanted. Stoved paints are at an advantage in a continuous factory line production where the shorter curing times are more compatible with production needs.

Aluminium building sheet is used extensively for cladding industrial buildings and office blocks. Of the total of 25,000 tonnes of sheet used in the United Kingdom, about 8,000 were painted, and this represents about 2 million square metres of surface requiring to be coated every year.

The largest application for coated extrusions is windows, with about 40,000 tonnes being used in the United Kingdom. Ten years ago they were nearly all anodised, but with the advent of electropainting and powder coating (see later), organic finishing has become predominant, particularly in the domestic field. Latterly, plastic windows have taken a share of the market.

8.1 COMPOSITION OF PAINT

A stoving paint consists of 3 parts:

(a) The Resin. This part is the binding agent, and gives the name to the type of paint. This provides the physical properties of the paint and decides its characteristics. A number are used for the manufacture of paint for aluminium and include alkyds, acrylics, polyesters, epoxies etc. The resins are complex organic compounds and are usually mixtures of several components. They are not unlike adhesives and provide the bonding between the resin and the aluminium as well as between the other constituents of the paint.

(b) The Filler. Titanium dioxide is the more usual filler or extender which is added to the resin to provide an opaque and dense appearance, but other ingredients are added also for different functions.

(c) The Pigment. This provides the colour, and consists of either organic compounds, such as dye, or they can be various coloured in organic salts or compounds of metals, such as copper or cadmium. Carbon is also used for producing black colours, or in combination with other pigments to produce dark shades.

The ability to resist fading by the action of sunlight is important for architectural paints, and, therefore, only pigments of known stability should be used. This may preclude some of the organic pigments which can fade when exposed to strong sunlight.

8.2 ADVANTAGES OF STOVED COATINGS

When painting is carried out in a factory, it is possible to control the conditions at all stages to provide a consistent quality. It also means that stoved or heat-cured coatings can be used with their improved properties compared to air-dried finishes. The advantages mainly conferred are that they are harder and more abrasion resistant, they have improved durability and chemical resistance and better adhesion to the aluminium.

The environment under which painting is carried out can also be monitored such as the humidity and freedom from airborne dust, factors which all contribute to the quality of the finish obtained.

Means of testing the finished product are also at hand, and one can, therefore, achieve a better product and longer lasting finish than by painting with an air-drying system on an open site under non-ideal conditions.

The use of a conveyorised and automatically-controlled paint line can improve the rate of production with economies in cost through lower requirements in manpower, and it also allows closer technical supervision.

Factory applied coatings are used for many components, such as domestic appliances etc., usually using stoved paint systems, but sometimes they may be oven assisted air-drying paints.

FIGURE 8.1. Aluminium boot lid for Rolls Royce Silver Shadow, stove painted with automobile finish processor after zinc phosphate pretreatment

Prepainting of architectural extrusions or sheet is increasingly carried out by the aluminium supplier, treated in standard sizes or lengths for subsequent cutting to size, drilling, mitreing etc. by the fabricator before erection. The painting of sheet for architectural purposes will be described later.

Much of the painting on aluminium has been carried out in the paint shops of fabricators, but more frequently use has been made of prepainted sheet or extrusions supplied by the aluminium producer. A large proportion is also treated by the jobbing paint applicator in the same way as anodising is carried out by anodisers. Toll painters have the advantage of flexibility, and should be more amenable to supplying shorter runs in a greater variety of colours or finishes than on the larger, automated plants associated with the in-house production unit.

8.3 APPLICATION OF STOVED FINISHES

It must be stressed yet again that the essential feature of any paint system is cleaning and pretreatment

These aspects have been extensively described in previous chapters.

Although etch primers have been applied as pretreatments for stoved finishes, they are not as effective, and are more costly than suitable chromate conversion coatings.

It is preferable that the pretreatment process is carried out at the place where the painting is to be carried out, but this is not always possible. Nevertheless, the treated aluminium should be dry and grease-free before coating.

8.3.1 Spray Painting of Extrusions

Painting of extrusions by application of wet paint by spraying, followed by stoving, is a frequently used method of treatment, particularly for coating aluminium used for architectural features. The most commonly used paint type is based on acrylics which provide a hard and durable coating capable of providing good adhesion to the aluminium. Other types of coatings such as alkyds or PVF_2 based resins are also applied.

The process has the advantage over the more recent electrophoretic painting and powder coating due to:

1. Lower cost of capital equipment in plant
2. Flexibility in colour change or paint type
3. Low cost for short runs.

Disadvantages are that the process may be more labour intensive and depends on the skill and conscientiousness of the operator. Because of this, it is more difficult to ensure consistent coating thickness within the controlled limits than on an automated line set up for a long run of similar components.

8.3.2 Cleaning Before Painting

Methods of cleaning have been described in Chapter 1, and it is an important stage that must not be neglected if a satisfactory standard of painting is to be achieved.

It is not sufficient to degrease by using solvent cleaning or by vapour degreasing. It is essential to accompany the cleaning with a degree of etching, either by one of the acid or alkali cleaners described. In the case of alkali cleaning, it is beneficial to precede the cleaning process by immersing the work in the acid used as a de-smut solution, and also to follow the cleaning stage by a similar dip. The first dip in acid is required with alloys containing magnesium, such as sheet in 5251 alloy or extrusions in 6063 and 6082. The reason for this is to remove any oxides of magnesium that may be present on the surface which are insoluble in alkali solutions and which would prevent uniform cleaning.

It is good practice when etching extrusions before painting to aim to remove 0.2 g/m^2 to achieve a clean active surface for the application of the subsequent conversion coating.

8.3.3 Pretreatment

Following etch cleaning it is necessary to treat the aluminium in a chromate conversion coating process by immersion or spray described in Chapter 1.

For maximum protection and adhesion, a hexavalent chromate conversion is recommended such as Alocrom 100, which will produce a green/blue colour rather than the gold coloured tri-valent chromium processes.

The aim is to produce a thin coating, with a coating weight of about 100 mg/m^2 rather than the thicker coatings used as protective or decorative finishes for sheet.

The coating should be allowed to dry, preferably at a temperature not more than 60–65°C. Subsequent painting should be carried out as soon as possible. It is bad practice to keep chromated work over a weekend, for example, before further processing.

It is preferable, also, to fix extruded aluminium lengths on to jigs to carry them through the various cleaning and pretreatment stages rather than to attempt to process the work in bundles. The latter practice can result in poor circulation of treatment chemicals during treatment and poor rinsing, causing patchy or stained work.

Two-pack etch-primers have been used where chromate conversion processes have not been available, but their use is not justified on a production flow line on account of their higher cost and inferior performance. A further disadvantage is the limited life of the primer after the two components have been mixed.

8.3.4 Application

Painting of extrusions is applied by spray guns. It is usual to suspend the extrusions horizontally on racks or jigs where upon they can be sprayed manually. Uniformity of coating depends on the skill of the operator as well as correct adjustment of equipment such as spray nozzle size, air pressure, paint viscosity, etc. Spraying must be carried out in a dust free atmosphere, the spray booth being equipped with air extraction to collect any overspray.

A clean filtered air supply is necessary, and a variety of spray nozzles, and gun settings such as air pressure etc. are necessary to control the finish achieved. Heated, air-less spraying can be used where the paint is applied to the gun under pressure from a pump to provide economy in paint usage.

More commonly, electrostatic paint spraying is carried out. Here the gun is charged from an electrostatic generator, and the aluminium is earthed. The droplets from the paint gun are given an electrostatic charge, and are projected forward and stick firmly in an even manner on to the aluminium surface. It also has the advantage of causing the paint to "wrap-round" extrusions allowing both sides to be coated at once. It also has the effect of reducing the paint usage by up to a half through lessening overspray.

Although racks of extrusions can be painted by hand-held guns before being loaded on to a conveyor, it is more frequent now for automatic paint-spray booths to be used. The aluminium extrusions loaded on to jigs, are carried through the spray-booth by a conveyor, where automatic electrostatic paint guns are oscillated vertically while the extrusions pass horizontally before them. Any overspray must be collected in the booth by exhaust extraction to remove the solvent gases and paint particles.

8.3.5 Curing

After painting the coated extrusions progress to the stoving oven, usually a tunnel oven with infrared heaters. The oven is divided into zones at different temperatures, each of which must be closely controlled. The first zone is kept at a lower temperature where the solvents are volatilised off. This must not be carried out too quickly to avoid "solvent-boil", a defect consisting of little bubbles on the painted surface. The aluminium then progresses into increasingly hotter zones to cause the aluminium to rise in temperature, the maximum being known as the "Metal Peaking Temperature". This is the temperature at which the curing of the paint is completed, which would have been determined by the paint manufacturer. It is important that this temperature is reached, as below it the chemical reactions in the paint causing cross-linking will not take place and the paint will be undercured with lower properties. Equally, the temperature must not be exceeded, or the paint may become embrittled and the colour changed. Regular calibration of the temperature recording instruments should be carried out to maintain the correct temperature gradient through the ovens.

The Metal Peaking Temperature for each paint is supplied by the paint manufacturer. The temperature achieved by the aluminium will be dependent on the size of the load; heavy and numerous sections will obviously require more heat. To determine correct operating conditions, the temperature of the aluminium as it progresses through the oven is usually measured with a thermocouple attached to the load and connected to a galvanometer.

Alternatively, the maximum temperature attained by the paint can be measured by use of thermal indicating paints or "Temple Sticks", which change colour irreversibly when a certain temperature is reached.

It is preferred that the temperature testing paint is applied to small samples of the same section being painted to ensure that the temperature of the samples will be the same as the main work. The test samples should also have been painted previously with the same colour that is being applied, since different colour finishes will absorb heat from the oven at different rates.

The temperature of an oven is controlled by altering the settings on one or more of the heating sections, but it is more usual to allow for different heat requirements of different loads by changing the conveyor speed.

8.4 SPRAY PAINTING OF SHEET

The principle of painting sheet by spraying is similar to treatment of extrusions. Again, a chromate pretreatment is preferred. Sheet can be painted by hand, where the articles are hung vertically and sprayed by air assisted guns, airless spray guns or by electrostatic methods, followed by stoving in a tunnel oven. This method is suitable either for fabrications made from folded sheet, or for flat sheets.

For the large scale production of flat sheet, or for treatment of profiled architectural sheet, it is more convenient to paint the pretreated sheet horizontally, the items being conveyed on a moving belt through the various treatment stages. Painting can be done under spray nozzles which reciprocate sideways as the sheet progresses at a controlled speed to achieve the required coating thickness. The sheet is then conveyed through the various stages of a tunnel oven.

Painting of sheet in this manner provides a uniform, consistent coating which can be closely controlled with minimum paint usage, particularly when applied by electrostatic means. It is useful in coating flat sheet, such as that required for container bodies and for transport applications, but its main advantage is its ability to treat sheet which has already been profiled, such as in the form commonly used for cladding buildings. By this method, it is possible to apply a hard and durable coat without having to compromise on flexibility as would be required if the forming were carried out after painting. It also ensures that weakness in the paint such as micro-cracking in the stressed areas of post-formed paint are avoided. The painted aluminium sheet, produced by British Alcan Aluminium, known as Baco Colourbond, is an example of a product that has found acceptance as one of the most durable architectural cladding sheet available in this field. This product is a combination of a resistant aluminium alloy, good pretreatment and selection after testing of a long lasting alkyd paint. (See Figure 8.2)

FIGURE 8.2. Cladding on CEGB power station. Stove painted after profiling of aluminium sheet — British Alcan Aluminium plc

Although the production rate for coating individual flat or preformed sheet can be slower than for continuously coating coil for subsequent forming, it can also prove more economical for smaller runs, since changes in the paint system or colour can be affected more rapidly.

The spraying operation can be carried out by conventional or electrostatic spray guns, but from the point of view of paint usage and efficiency of application, it has proved to be more economical to apply the paint with an electrostatic blade coater.

In this application, the paint is allowed to run over an electrostatically charged blade, held over the earthed sheet as it passes underneath. The charged paint particles are propelled on to the sheet in a uniform pattern with little wastage through overspray.

8.4.1 Coil Coating

Improvements in paint and the increasing demand for colour has brought about a considerably increasing need for painted aluminium sheet. As automation progresses, it has become more usual for the manufacturer to turn to the use of aluminium sheet which has already been painted, and to close down his small and often inefficient individual paint shops.

Improved paint formulation has provided flexible paint coatings which

FIGURE 8.3. Aluminium road sign, made from blank, chromate treated and stove painted on reverse side. Retroreflective film on front face

allow a fair amount of forming to be carried out without damaging the paint coating. Use of the more formable coatings can allow pressing operations on prepainted sheet as well as the more simple bending. Improvements in the use of adhesives overcome problems of jointing components using prepainted aluminium where spot-welding was not always possible.

These market opportunities have opened the way to the production of large quantities of prepainted sheet and coil which can be used in the architectural and industrial fields.

To satisfy the demand, coil coating of aluminium has been carried out in the United Kingdom since about 1960, and now European plants capable of treating wide coil up to 2 metres wide and over are in operation. These plants treat large coils of aluminium at speeds of up to 100 m/min, depending on gauge. (Figure 8.4).

The lacquering of thin gauge aluminium used for containers in the Packaging Industry takes place at even higher speeds.

The coated coil is then ready for further operations such as press-braking or roll-forming into profiled sheet used in the building or caravan industries. Painted coil with flexible coatings can be used for the manufacture of shallow drawn components

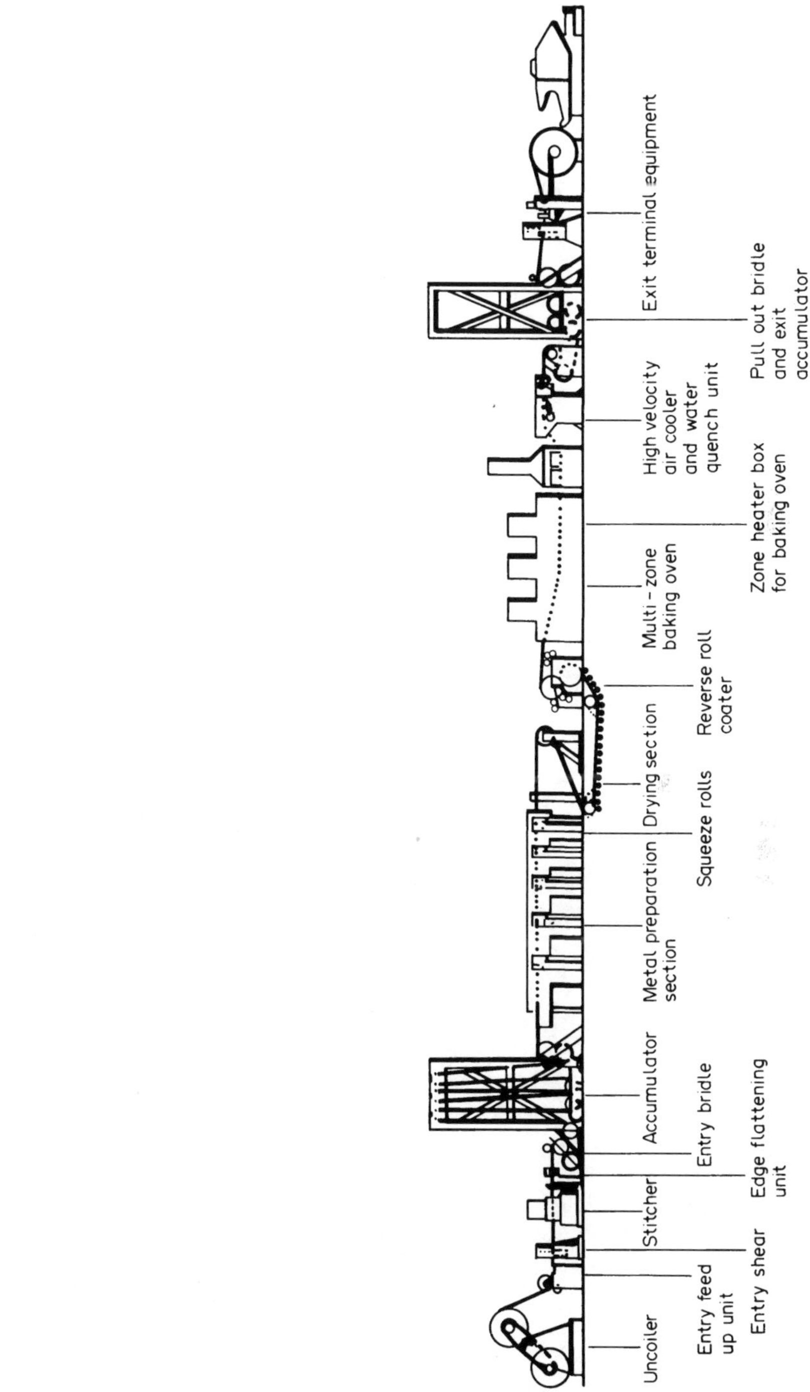

FIGURE 8.4. Diagram of a typical strip-coating line for treating aluminium coils

8.4.2 The Plant

A paint line capable of treating wide coil is obviously capital intensive, but is able to produce process aluminium at a high rate with low labour requirements. The plant can coat aluminium coil or strip on one or both sides, being fed in from coil at one end and is drawn out under tension at the other after treatment through the various stages. These are as follows:

1. de-coiler
2. stitcher
3. accumulator
4. de-grease
5. pretreat
6. coat
7. stove
8. accumulator
9. re-coil

The aluminium coil is loaded on to the entry end of the line where the front of the incoming coil is joined on to the tail of the previous coil. This can be achieved by spot welding, but is more often carried out by a mechanical stitching method. The coil then passes into the accumulator which simply consists of a system of rising and falling pulleys which allow storage of the strip.

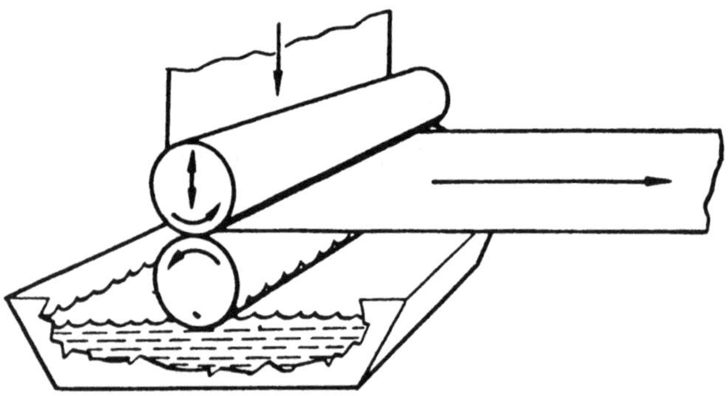

FIGURE 8.5. Reverse roller coater

The stitching operation may thus be carried out on strip which is locally stationary without the need to stop the main part of the line.

The first stage of treatment is cleaning, carried out in an enclosed chamber, where hot alkaline cleaners are sprayed on to the surface by pressure jets to remove any contaminants and slightly etch the surface. This is followed by spray washing followed by an acid de-smut solution, again by spray jets.

The aluminium then undergoes a chromate conversion coating followed by rinsing in two stages, the latter being with de-ionised water before drying by wiping and then by evaporation in hot air. Recent developments permit the conversion coating process to be followed by drying without rinsing.

The liquid paint is applied by a reverse roller coater which is adjusted to provide a uniform and closely controlled coating thickness over the width of the sheet (Figure 8.5). One or both surfaces can be coated with paint of the same or different colour.

If paint is applied to one side only, the reverse side is coated with a clear lacquer to facilitate lubrication of the surface during coiling or subsequent post forming operations.

Finally, the coated aluminium passes through the various drying and stoving zones of the oven. The coil, after cooling, enters into a second accumulator before being re-coiled, or levelled and guillotined into individual sheets.

FIGURE 8.6. Touring caravan manufactured by Cosalt Caravans Ltd from coil-coated aluminium — Euramax Aluminium Ltd

Naturally, coil coating is more suitable for longer runs than can be carried out by conventional coating lines, as it is uneconomical to change frequently the paint to be applied. It may take about 4 hours to complete a paint or colour change on a line.

The advantages are:

— high throughput

— a good uniform surface appearance
— a finished coating of consistent quality
— close control of product.

With a large amount of metal being processed at one time, processing failures can be expensive, so close control of cleaning and treatment chemicals are necessary as well as the requirements of paint application and the quality of the paint itself. However, given adequate monitoring of these requirements, a consistent and high standard for the product is achieved.

8.4.3 Paint Types

The types of paint applied in coil coating have been to a large extent based on acrylics. These can be given good forming properties and have good weathering resistance, enabling this product to be widely used for architectural applications, and for coating sheet used in the caravan, road transport and domestic appliance industries.

FIGURE 8.7. Cladding on factory building made from coil-coated aluminium which was subsequently profiled — Euramax Aluminium Ltd

When used for profiled building sheet, painted coil is required to have a coating which is sufficiently elastic to stand up to the forming operations of the sheet.

Nevertheless, the paint coating is stretched at convex bends and even the more flexible acrylic coatings can suffer from micro-cracking causing sites of potential weakness. For very long lasting requirements, paints based on polyvinylidene fluorides (PVF_2) resins are available. This type of paint is more flexible than the acrylics on their own, and will provide a longer lasting protection. This type of paint is more resistant to chalking on exposure on the outside of buildings. This is a defect due to gradual breakdown of most paints with the formation of a slightly chalky deposit on the surface, which has the effect of altering the appearance of the painted surface.

A PVF_2 paint coating is more difficult to apply than other paint coatings, requiring an undercoat or primer as well as the top coat to provide good adhesion. It also requires different stoving conditions compared to other paints. Although the paint itself is more expensive, the advantages of good flexibility, low chalking rate and long life are worth considering for many applications.

Vinyl paint systems and organosols have good forming characteristics, and are used for coil that is going to be used in the manufacture of pressings. They are often applied to aluminium foil when intended for the packaging market. Vinyls are not suited for outdoor exposure since they have low chalking resistance and the colours can fade.

Stoved epoxy coatings are commonly used for providing chemically resistant coatings in the food industry, and in areas where good resistance from acid of alkali bearing atmospheres is required. These coatings tend to be hard and have good adhesion resistance, consequently they have been used for aluminium furniture and domestic appliances. In spite of their ability to provide good protection from acids and alkalis, they are not suitable for use in exterior environments since they are more prone to chalking defects, brought about by the action of sunlight.

A recently developed coating is applied to coil, where the paint is reinforced by fine nylon particles resulting in an abrasion resistant coating which is formable and only slightly less durable than the PVF_2 finishes. It is produced under the name Duralcoat 80 by Alcan Plate Ltd.

Chapter 9

Electropainting of Aluminium

9.1 THE PROCESS

Paint application by electrophoresis was originally developed for coating steel and other metals, and it was only comparatively recently that a description of the process as used for treating aluminium was published. The first commercial plant was installed in the mid-1960s in the United States for coating aluminium extrusions, since when many other plants have been built. In Britain, the first plant was installed about 10 years later, followed by a number of others.

The method of operation has features similar in some ways to anodising. The basic principle is to suspend the work to be coated in a water-based paint suspension, and to apply a voltage, whereby charged paint particles are attracted to the aluminium and adhere firmly to the surface. The adhesive forces are sufficient to allow the work to be withdrawn from the bath and washed free of excess paint; the adherent paint coating is finally cured by stoving in an oven. Since no organic solvents are used, no toxic emissions are generated on curing.

As with anodising, the work is required to be fixed on to aluminium jigs or racks, to allow it to be transported through the various process stages, and to permit electrical contact at the paint stage for application of the electric charge. The sequence to be followed consists of these stages:

1. Etch clean
2. De-smut
3. Electropaint
4. Dry and stove.

Thorough rinsing and washing is necessary after the different treatment stages to avoid contamination.

The process has been adopted for the painting of long extruded lengths of aluminium which are to be fabricated in subsequent operations into architectural features, in particular for aluminium primary or secondary windows, and for double glazing.

This method of processing is particularly suitable for the large-scale output required for this market, since process times are short, i.e. 4 minutes for paint deposition, thus offering opportunity for treatment on a fully automated line. A typical plant would consist of about 10 tanks containing around 2000 gallons capacity, each about 8 metres long, capable of treating 7 metre lengths, and treating about 70 tons of sections per week.

The advantages of this method of painting aluminium by electropainting are:

1. Uniform paint coverage. The uniformity of the paint coating is high, and good coverage of the aluminium is achieved. Coverage of sharp edges is good compared with other coating methods, and the "throwing power" is high, which results in penetration of recesses to achieve a uniform coating.
2. High output rate.
3. Controllable coating thickness.
4. No paint loss. Excess paint washed from the aluminium is reclaimed for re-use by filtration.
5. No effluents or fire risk from evaporating solvents.
6. No chromium-containing pretreatments necessary with subsequent effluent problems.
7. High standard of surface finish.
8. Extremely good adhesion between paint and aluminium.

The disadvantages of the process compared to other methods are:

1. High capital cost of the plant.
2. Limited maximum thickness of coating.
3. Necessity for careful control of production and composition of paint.
4. Lack of flexibility with regard to colour or paint type.

Recent improvements have enabled coatings up to 40 μm thick to be applied compared to the earlier 25 μm maximum.

9.2 PRINCIPLES OF ELECTROPAINTING

The most commonly used process for electropainting aluminium is anodic, although cathodic systems have been used. This is in contrast to electropainting steel where, although initially anodic processes were used, the trend is now to use cathodic processes which provide better protection for the steel. This is frequently used in the car industry for electropriming steel components.

In the case of aluminium, anodic processes have the main advantage of requiring none of the usual pretreatment coatings based on chromates or phosphates which are necessary for conventional painting. With electro-painting, since the aluminium is made the positive electrode or anode, a thin but finite anodic coating is formed on the surface of the aluminium as the paint deposition progresses which provides an extremely good key to the deposited paint coating.

The "paints" used for the anodic process are specially formulated dispersions of paint solids in water, consisting of resins, pigments and other components. When first made up as a suspension, they are not soluble in the water, but they become partially hydrolysed on treatment with alkali such as ammonia or caustic soda when required for use.

$$\underset{\text{insoluble}}{RCOOH} + OH' \rightarrow \underset{\text{soluble}}{RCOO'} + H_2O$$

By controlling the pH of the suspension with an alkali, the paint particles become partially soluble as negatively charged groups, and in the paint tank they will migrate to the positively charged aluminium, where they will be neutralised and become once more insoluble. At the same time they will be deposited on the aluminium to build up a coherent and solid paint coating layer.

The paint particles will first be deposited on to the nearest or easiest parts of the surfaces of the aluminium to be presented, but as the paint layer builds up, the electrical resistance increases, forcing the inaccessible areas to be coated. Thus the process is self-regulating in so far as an even and uniform coating will be formed even on those areas which would be difficult to coat by other means.

The rate of deposition depends on the applied voltage, and the thickness that can be achieved will thus depend on the voltage as well as time and temperature.

As painting proceeds, ammonia is liberated at the cathode and is lost. If this did not occur, the pH would rise and the paint ions would be rendered insoluble once more, so to some extent the control of pH is dependent on this loss of ammonia. The process requires the additions of amines to preserve the balance.

One method patented by ICI is to use potassium hydroxide to control the pH and to surround the cathodes with a semi-permeable membrane which would allow passage of potassium ions to the cathode, but would retain the potassium hydroxide formed there which can be removed by washing out with water. Since this system would only work when painting was in progress, an ion exchange resin system is incorporated in the membrane to prevent the potassium hydroxide from diffusing back into the main paint suspension when the voltage is not being applied.

The advantages of using a membrane to regulate the pH are that control is simpler, the paint system is more stable and no pre-mixing or conditioning of paint has to be made in a separate tank.

The paint solids are maintained at 10%, and since paint requirement is dependent on ampere hours usage, a coulometer can be used to regulate routine paint additions.

9.3 PROCESS DETAILS

Although hand-operated plants for treating small components would be feasible, commercial electropaint plants usually consist of lines of tanks adjacent to each other, which are served by cranes transporting work through the stages.

The cranes can be operated by hand or the whole line can be completely automated allowing work to be carried through each operation under preset conditions. This type of plant is suitable for pre-painting of lineals which are to have subsequent work carried out on them, such as sawing to length, mitreing, drilling etc. As with most anodising plants, the extrusions are processed at or near to the horizontal.

One plant in the United Kingdom treats pre-formed extrusions in short lengths, vertically. The work is carried through the tanks by a conveyor, and being vertical, washing and drainage after each stage is more easily and efficiently carried out.

9.3.1 Jigging

The work to be painted must be firmly fixed to aluminium jigs to provide passage of the electric current to the work and to allow the work to be transported through the various treatment stages. Jigs are made with vertical spines provided with spring clips to hold the aluminium extrusions horizontally (see Figure 9.1). As with anodising, good electrical contact must be maintained and this requires any remaining paint coatings on the contact points to be removed, generally by filing. After about 15 cycles, the jigs require to be stripped of deposited paint, usually by immersion overnight in a cold solvent paint stripper.

Extrusions are usually jigged near to the horizontal, or with a slight slope which will improve drainage after each stage, and care is needed to avoid areas where air may be entrapped which would prevent deposition of paint. One plant in the United Kingdom hangs the extrusions vertically which allows good drainage after each stage, and the work pieces are carried through the line on a continuous conveyor (Figure 9.2).

9.3.2 Pretreatment

The first stage is to provide good etch-cleaning of the aluminium, usually by one of the alkaline cleaners described in Chapter 2. It is important to

FIGURE 9.1. Aluminium extrusions being processed through electrocoating line — Electropaint Ltd

etch a finite amount of aluminium, usually about 0.2 mg/cm^2 to provide a clean surface. After washing, the smut that may be formed is removed by an immersion in 10–20% nitric acid. After one minute, the aluminium is removed and washed in two stages, first in a tank of running water, and then by either a second rinse of clean running water, or in a static de-ionised water tank which is being continuously passed through the de-ioniser to maintain its purity. As work is removed from this second rinse, a spray of deionised water is directed on to the aluminium as it leaves the water. Thorough removal of all contaminants is essential to preserve the life of the paint tank.

9.3.3 Paint Application

The cleaned and etched aluminium is lowered into the paint suspension. A voltage is applied from a rectifier at about 160 volts to provide a constant current density of about 0.1 amps/dm^2. The paint systems used for aluminium require the metal to be made the anode and, during the deposition,

FIGURE 9.2. Aluminium extrusions suspended vertically on a continuous conveyor entering paint bath for electrophoretic coating — Electropaint Ltd

a thin anodic coating is formed of about 2000 Å (0.2 μm) at 160 V. This is sufficient to form a strong bond between the aluminium and the stoved paint without the need for a normal pretreatment such as a conversion coating process as required by other coating processes.

The tank containing the paint suspension can be made of mild steel with either a plastic or epoxy coated lining. Stainless steel or aluminium cathodes are required, the area and position of which are important with regard to the uniformity of the paint coating.

The paint is a water-based suspension of 10% solids which require agitation by filtered air. The paint suspension is continuously circulated by pumping from a holding tank which is monitored to control the level of solids and the pH. The paint must be continuously circulated with a paint turn-over of about 4–5 times per hour, even when not in use, to prevent settling out of solids. The water used in the paint make-up and for topping up must be de-ionised to prevent the build up of solids or contaminants which would cause problems.

The operating temperature of the paint tank should be maintained at 30°C, as coating thickness will vary with temperature. Since heat is generated by the power used during painting, cooling coils may be necessary to control the temperature.

FIGURE 9.3. Aluminium window using sections prepainted by electrocoating in white acrylic

Electrical power for the painting process is supplied by dc rectifiers. The work is loaded into the bath with the supply "off", and the power is applied gradually to prevent a surge of current.

If the process is to be operated at a constant current density rather than at a constant voltage, it is desirable to know the surface area of the load that is being treated. This requirement is similar with anodising which is normally carried out at a constant current density. When the necessary current applied to the load is achieved, it will be necessary to raise the voltage in stages to maintain the current density, since the resistance of the anodic coating and the paint coating will rise as deposition continues.

Some paint systems can be operated at a constant voltage throughout the cycle.

The amount of paint deposited will be largely dependent on the number of coulombs used, but it is also dependent on the conditions of the paint bath. The rate of increase in voltage to maintain the correct current density can be determined for a particular set of conditions of the bath, and can be controlled automatically. The coating thickness achieved will depend on the number of coulombs used per unit of area, and a coulometer can be used to control the coating thickness as well as the rate of paint usage.

The dc voltage required to be applied is high compared with anodising, and will depend on the paint system used and on operating conditions. It could be as low as 120 volts, but some systems require up to 300 volts. Although the current density at up to 0.25 amps/dm^2 is lower than for a conventional anodising, the power required is higher (although applied for a considerably shorter time), hence the need for adequate cooling of the bath.

Because of the high voltage employed, adequate electrical shielding is required around the tank, with automatic cut-off equipment, to prevent accidents from electrical shocks.

A processing time of around 6–7 minutes is usually necessary to deposit a coating thickness in the order of 25 μm, the standard for architectural applications, although 40 μm coatings are now possible.

9.3.4 Rinsing and Ultrafiltration

After removal from the paint tank, work is transferred to a de-ionised rinse tank to remove any surplus or non-adherent paint. The water from the contaminated rinse tank is fed continuously under pressure through an ultrafiltration plant, to collect any residues of paint which are then returned for further use. A certain amount of the filtrate is bled off to waste to prevent a build up of any soluble contaminants, and the remainder is fed back into the paint tanks through spray jets over the tank to assist in washing off surplus paint from the coated aluminium as it leaves the paint.

The system of a closed loop de-ionised water rinse before work enters the paint prevents any contamination of the paint bath, and the ultrafiltration of residues of the rinse tank after painting ensures very high recovery, and a well-run paint line should produce a paint efficiency of 95%. It also ensures the water usage of a plant is kept low.

9.3.5 Stoving

The final stage is the curing of the paint. The jigs loaded with the coated aluminium are loaded on to a moving conveyor which carries the work through a camel-back oven. The correct temperature is necessary to achieve adequate curing but without the danger of over-baking. Control is achieved by a combination of conveyor speed and oven temperature, the criterion being the maximum temperature which the painted metal will attain (Metal Peaking Temperature), and for how long. The necessary data is available from the paint manufacturer, and the correct conditions are controlled on the plant using a thermocouple to measure metal temperature.

9.4 COSTS

The capital cost of an electrocoating line would be similar to that of a similar sized anodising plant producing lineals for the architectural market. It would be slightly higher than a similar plant to apply paint by electrospraying or powder coating. Etching, cleaning and stoving requirements are similar, but unlike powder coating or conventional liquid painting electrocoating does not require a chromate conversion coating treatment with its attendant effluent problems. Operating costs are similar to those of an anodising plant.

A typical plant, capable of treating 7 metre lengths, would occupy about 650 m^2, would employ about 10 men per shift with an output of 75 tonnes per week. A high output can thus be achieved with a consistent quality.

9.5 CONCLUSION

The stoved finishes achieved by electropainting result in hard, durable coatings with good appearance, and are ideal for the pre-painting of extruded aluminium intended for subsequent fabrication of aluminium window sections and other architectural use. The paint quality is high, and when correctly applied and stoved, will satisfy the requirements necessary for the exteriors of buildings. Acrylics are commonly used as well as polyurethanes, although other resins could be applied. Apart from the plant which treats short prefabricated lengths vertically on a continuous conveyor, the system is ideal for treating horizontal lengths up to 7 metres long.

9.6 DEVELOPMENTS

Although the market for painting extrusions, particularly for domestic housing, mainly requires a white colour, a difficulty is the inability to provide other colours owing to the high capital costs of installing a separate paint tank for every colour. Nevertheless, there is one plant which is also able to apply paint in a dark brown colour.

An interesting application for electropainting of coiled aluminium strip has been suggested. This is coating coil with an insulating lacquer which could be used for electrical winding strip for transformers and coils. An advantage is the extremely good edge coverage achieved by electropainting.

A disadvantage of electropainting, compared with powder coating, has been that the coating thickness was initially limited to 25 μm but now 40 μm maximum is possible. Although this thickness satisfactorily meets most of the required standards, the coating has to withstand damage from mishandling and from fabrication treatments before erection. Powder coatings, on the other hand, are applied with coatings ranging from 50 to 100 μm, which are more able to withstand any mechanical handling damage which is a considerable advantage.

9.7 THE FUTURE

The present process is dependent on the anodic coating formed at the same time as paint deposition to provide the necessary adhesion between paint and aluminium, and this has proved effective. In Japan, a chromate conversion process is applied before painting, as it is claimed that the chromate provides increased corrosion resistance to the painted surface. Further developments may lie in improved paints for longer life and in increasing the thickness of coatings that can be applied.

Chapter 10

Powder Coating of Aluminium

Powder coating of aluminium, extrusions in particular, has now become an important feature in the finishing of aluminium. The process was largely developed in Europe, and has since been used in the decoration and protection of aluminium used on many buildings, as well as in many other applications.

The main application has been in the architectural field where colour is required as an alternative to the more functional appearance of the natural finish of plain anodised aluminium. Colour anodising has been for the main part limited to the production of the very stable bronze gold and black colours produced by integral or electrolytic colouring processes (see *The Finishing of Aluminium and its Alloys* by V. F. Henley). The advent of stable and resistant coloured organic finishes has allowed more scope for the designer and architect to provide a wide range of decorative coatings with a life approaching that of the anodic finishes.

The development of reliable and long lasting coating systems by the applications of fused powders was encouraged by the rising cost of solvents used for conventional painting, partly brought about by the oil crises of the 1970s, and the attraction of solvent-free coating processes where dangers from fire and atmospheric pollution have to be avoided. Since it was a new field requiring fresh technology, it perhaps attracted a more scientific approach to decorating aluminium than had previously been carried out by the rather empirical exponents in the field of spray painting.

The process has a similarity to the electrostatic spraying of paint. Rather than applying charged droplets of liquid paint, small solid dry particles of "paint" are charged electrostatically, and propelled by the charge on to the earthed aluminium. The powder builds up evenly on the surface to which it is firmly held, and finally the powder is fused under heat into a continuous layer. The reaction is more than simple fusion, as the different constituents making up the powder are made to react with each other and with the underlying aluminium conversion coating to form a coherent and firmly bonded coating.

Particle size and distribution are important to provide even coverage of the finished coating, the size generally being in the range 20–120 μm.

Epoxy resin-based powder coatings have been used for providing hard abrasion-resistant finishes, particularly suitable for furniture and for other indoor and wear-resistant applications. They are not suitable for exterior applications since under conditions of weathering they are prone to chalking. This is a powdery coating formed on the surface as a result of breakdown of the resins by sunlight which also produces a change of colour. Epoxy powder coatings are used where good corrosion protection is required, particularly in acid environments, and for protection of pipelines.

Various resins have been used in the manufacture of powder coatings for aluminium to be used for exteriors of buildings, and these include acrylics, polyesters and polyurethanes. Of these, specially formulated polyesters are finding increasing use in the architectural field, due to their ability to retain gloss, their resistance to chalking, and, when made up with suitable pigments, their good colour stability. The success of polyester resins has, to a large extent, been due to the testing and marketing of polyester powder coatings known as Syntha Pulvin made by Vereinigte Pulverlack GmbH of Germany and now made and marketed in the United Kingdom by Syntha Pulvin Ltd. in Birmingham.

The success it has gained in the architectural field has not only been due to careful testing of the product and effective marketing, but also to the efforts made by the manufacturer to attempt to monitor the application of the powder by ensuring that the users of the powder follow correctly the laid down processing procedures. Many other powder manufacturers now produce similar polyesters for this application, also of a good standard, where colour stability and durability are achieved.

10.1 METHOD OF APPLICATION

The application of powder coatings to aluminium is carried out in a similar sequence to that for wet spraying. The aluminium requires to be cleaned thoroughly as for all applied coatings, and an effective and efficient chromate conversion coating process must be carried out. The effectiveness of the subsequent coating depends to a large extent on the degree of bonding, and this is best achieved by the green coloured hexavalent chromium conversion coating processes such as ICI's Alocrom 100 and other similar products. Other types of pretreatment methods have been tried, such as phosphate treatments and even unsealed anodic coatings, but the bonding achieved is less effective. Zinc phosphate pretreatments are useful in that

FIGURE 10.1. Hand-held guns and equipment for application of electrostatic powder coating — Mindon Engineering Ltd

both steel and aluminium can be treated in a mixed assembly, but where there is a choice, chromate coatings are to be preferred.

The powders used for coating are specially prepared by fusing and blending suitable resins and other additives with selected pigments, grinding, and finally sieving. The quality of the finish that is achieved depends on the particle size of the powder, that the size range is carefully controlled and the powder must be kept free of contaminants.

Specialised equipment is necessary for application of the powder. The powder is applied by an electrostatic gun which charges each particle from a generator. The particles are then propelled on to the aluminium to be coated, which is earthed, and are held firmly in place on the surface.

By the nature of the process, there is a considerable "wrap-round" effect, and the powder will, to some extent, coat all sides of the object, although to achieve the best uniformity of coating thickness, the gun should be made to traverse all faces.

As with wet painting by electrostatic spray, there can be problems in ensuring even penetration of narrow re-entrants with extrusions (the Fara-

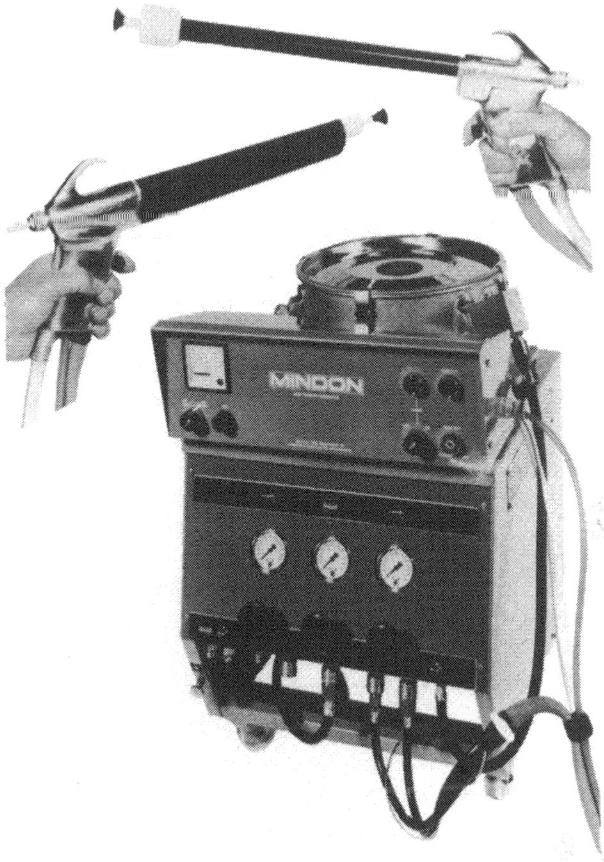

FIGURE 10.2. Application of powder coating showing spray booth, gun and generator. Background shows cyclone for reclamation of powder from overspray — Mindon Engineering Ltd

day cage effect) which prevents deposition of the charged particles in these areas. Nevertheless, by correct manipulation of the spray guns and selection of voltage it is possible to coat intricate surface areas with a uniform coating over most of the surface.

The operation of powder coating can be carried out in a small hand-operated plant, using a hand-held gun with the objects positioned in a spray booth, followed by stoving in a batch oven. In a large plant for treating extrusions, the process can be automated with the work on a conveyor and using automatic guns, tunnel spray booths and ovens. It is now possible to powder coat extrusions at rates approaching the high outputs of electropaint plants.

10.2 EQUIPMENT

10.2.1 Electrostatic Generator and Gun

The principle of powder application is to charge each powder particle as it leaves the spray gun so that it becomes ionised. It is projected to the earthed aluminium to be coated where it is held by electrostatic forces. As the coating builds up, the resistance increases and causes the following charged particles to become attracted to less resistant areas. Thus, the coating is to some extent self-regulating, with a uniform thickness on all faces, not only on those directly in line with the gun.

In order to charge the powder particles, a high voltage dc generator is required with an output voltage of about 100 kV. The voltage is applied by a heavily insulated high tension cable from the generator to a rotating disc at the front of the spray-gun. Dry air is fed into the gun and by a venturi nozzle, powder is drawn from its container and blown on to the disc. There the particles become charged, and are propelled on to the earthed aluminium. In some systems, the charge is generated in the gun itself. The efficiency of deposition depends on:

The electrostatic field
The ionisation current
The resistivity of the sprayed material.

The rate of powder deposition which in turn affects the coating thickness is dependent on the air pressure, which is used to control the spray pattern from the gun, and the powder flow to the atomiser in the gun. The voltage controls the charging of the particles and the ionisation current, which can also be affected by the resistivity of the power particles.

A small hand-held gun can have a small powder container incorporated on top of the gun, to allow the powder to be fed by gravity, but the powder can also be fed from a separate container by a mechanical screw feed, or by an air venturi from the powder which is held in suspension in a fluidised bed.

With the high voltages employed for the charging of the powder particles, a necessary high standard of electrical insulation is required for the supply cable from the generator to the gun and for the gun itself. It is also desirable to have safety devices to cut off the power supply instantly should the gun be earthed accidentally during use.

The air supply for the supplying of guns need not be of a high pressure, but must be oil-free and dry, which can be achieved by filtration and freeze-drying.

10.2.2 Powder Recovery

Although the proportion of powder deposited on the work is around 50%, most of the excess powder can be recovered by a suitable spray booth, equipped with cyclone separators for entrapping the larger particles. The proportion of very fine particles are more difficult to remove, but are recovered by cartridge filters.

The recovered powder can be used again and, after sieving, is usually mixed with fresh powder for re-application. With an efficient cyclone and filtration system, it is possible to recover over 95% of the powder for re-use, making the overall recovery of the process very efficient.

The safety aspect of the spray guns has been dealt with. A possible hazard with powders is the danger of explosion should a high powder/air ratio be ignited by a spark. The danger is avoided by providing excess air drawn into the spray booth by the exhaust fans and to ensure the powder content is kept below 10%. The material from which the booths are made is often itself powder coated to insulate the surfaces and prevent adhesion of charged particles. Booths made from plastic panels have also been suggested for the same reason.

Current health and safety requirements and factory inspectorate regulations require explosion relief panels to be fitted which in the event of an explosion would act as a safety valve and fail in preference to other areas which might cause more damage. In practice, the possibility of an explosion occurring is very slight, and the fire risk is less than with a solvent used in a liquid paint plant.

10.2.3 Stoving

In common with other thermosetting systems, the powder coated metal has to be stoved to cure the resins. Not only must the temperature be reached to achieve curing, but the temperature of the metal and coating must be high enough to complete the cross-linking of the resins, and the particles must be fused to form a smooth and continuous layer.

The heating requirements are different from stoving of solvent-based paints. There is no need for a low initial application of heat as there are no solvents to be evaporated, and the full heating can be applied immediately.

The range of temperatures and time to achieve full curing depend on the type of powder coating, but a typical polyester powder coating requires from 10 minutes at 210°C to 25 minutes at 190°C. Too low a temperature results in poor adhesion and low durability whilst too high a temperature causes yellowing and embrittlement of the coating.

10.3 POWDER COATING PRODUCTION

The production of powder coated aluminium depends on the size, quantity of the aluminium components and the type and colour of the coatings to be applied.

This will depend on the equipment used. The process can be carried out on a small hand-operated plant with or without varying degrees of conveyorisation, or if production requirements are large, fully automated and conveyorised plant is suitable.

As with conventional wet spraying of paint, the usual cleaning, etching and conversion coating pretreatments are necessary. Where possible, a chromate conversion process is preferred for maximum adhesion.

To ensure satisfactory all-round surface contact with various solutions, work pieces are preferably mounted individually on jigs for transporting through the different process tanks. The jigs may not necessarily be the same as those used to hold the work when being coated, although on an automatic plant they would be.

The final rinse after treatment should preferably be by de-ionised water to avoid contamination of the surface, and the work should be dried at a temperature not greater than 60°C. The treated aluminium should be powder coated as soon as it is dry, and it is bad practice to allow the aluminium to remain overnight before coating.

In most small plants, it will be found more convenient to pretreat in a series of tanks adjacent to each other when each rack load of work is individually immersed in each tank, either by hand for small components, or by an overhead crane for treating large loads of extruded lengths.

A number of automatic lines are designed with a continuous conveyor on to which the jigs are loaded for transporting through the different process stages. The pretreatment stages are then carried out by the work passing through a tunnel divided into stages where the various solutions are directed on to the work by pressure jets or sprays. Providing the sprays are correctly designed to ensure total coverage of the work, more concentrated solutions can be used and the effectiveness of the treatment compared to a dip process is improved thereby enabling process times to be reduced. The method also allows greater control and a more effective treatment since solution is being continuously directed on to the aluminium from a holding tank which can be continuously monitored to keep it in balance.

After rinsing and drying, the work is ready to be powder coated. On a continuous line, the jigs pass through the spray booth which has one or more spray guns on one or both sides. The guns oscillate vertically from top to bottom depositing powder on all faces. Finally, the coated aluminium passes through a tunnel oven where it is cured, and finally cooled.

With a small hand-operated plant, the loaded jigs can be loaded on to a conveyor carrying the work through the coating and curing stages, or they can be mounted in front of a spray booth individually, sprayed by hand and finally loaded into a batch oven.

10.4 COLOUR

The powder coating of aluminium has been dominated by the aluminium window market, for either primary windows, secondary windows (double-glazing) and replacement windows, which have largely required a white finish. Probably only 10% of this market is required in colour. Thus, all the larger powder-coating lines have geared their production to white. Nevertheless, powders in many colours are available which are lightfast and can be used for outside situations. Provision of colour by powder coating is, therefore, feasible, but at a price due to the time taken in cleaning or changing over spray guns and booths. This can add 10% to the cost of applying a colour. From this point of view, powder coating is intermediate between electrocoating lines which are inflexible, and lines applying liquid finishes, which allow for a more rapid colour change without hold up.

The problem is the need to clean down guns, spray lines, feeders and spray booths with their ancillary equipment of all residues of powder before embarking on a new one. Considerable effort has been spent by plant makers in designing plants to reduce this unproductive time interval, nevertheless, coloured coatings bear a higher price to compensate for this. Where production rates are important, this can be overcome by having two spray booths. On automatic conveyorised lines, the spray booths can be mounted on wheels so that for colour changes, the alternative booth is moved into the line.

In order to standardise the colours available, the policy has been, even for British manufacturers, to follow the German RAL series for colour, since it was largely in that country that powders suitable for buildings were developed.

The use of thermosetting powders for many applications is likely to grow. The main changes being brought about are likely to be in the powder itself to reduce stoving times and temperatures. The problem at present is that the stoved finish is more likely to suffer from the orange peel defect, which is a function of the particle size of the powders, the coating thickness and their physical properties. Methods of applying the powder to improve the uniformity of coating thickness and to reduce the average thickness to levels around 50 μm will provide an economy in raw materials but would still provide good protection for mechanical damage.

10.5 THERMOPLASTIC POWDER COATINGS

This group of coatings depends on heat fusing the particles and causing them to flow; they do not undergo a chemical change as with the thermosetting group. These types of coatings have been used for a considerable time, and include in particular, pvc and nylon.

These powders can be applied by spraying on to preheated aluminium, but very often the fluidised bed technique is used. With this method powder is contained in a vessel and air is diffused through the base which mixes with the powder which is then suspended in the rising air to form a volume about three times greater than the compact powder. This acts in part as a liquid, and when heated objects are lowered into this bed, the particles surround the components and adhere to the surface.

If the components are thick enough, they may have sufficient heat content to cause the particles to melt to form a fused layer. Otherwise for thin gauge components, the final fusing can be completed in a separate oven.

Coatings are considerably thicker than those produced by electrostatic powder coating, and are mainly used for protection rather than for decoration. They are not as hard as thermosetting coatings, but do have good impact and chip resistance.

10.5.1 Polyethylene (Polythene)

Coatings are used on aluminium to provide chemical protection. They are used for protecting aluminium from corrosion when it is embedded in soil or in concrete.

10.5.2 PVC

This coating provides good chemical resistance and has been used to coat aluminium jigs used in the anodising process to prolong their life on immersion in alkali etching solution.

10.5.3 Nylon

Coatings provide extremely good chemical resistance, although they are more expensive than other thermoplastic finishes. They require a primer to ensure good adhesion to the aluminium, and are suitable as coatings that are to come into contact with foodstuffs.

10.6 FACTORS AFFECTING USE OF THERMOPLASTIC COATINGS

An advantage in the use of thermoplastic coatings is that slightly underheating or overheating does not harm them in the same way as for the thermosetting group. Plant equipment and application are simpler and this method is very suitable for treating small components. The powders used are more stable on storage, and there is no loss on application.

Thick coatings can be obtained, but the disadvantages are that it is difficult to control the thickness within limits, and edge coverage at corners is difficult to achieve because of the high viscosity of the coating when molten.

The process depends on its effectiveness by completely enveloping the coatings to be coated with a thick skin and only effective de-greasing is necessary. For more demanding applications where adhesion is important, it is beneficial to apply a conversion coating to the aluminium before coating.

Chapter 11

Selection, Maintenance and Restoration of Organic Coatings for Architectural Applications

11.1 INTRODUCTION

The various methods commonly used for decorating architectural aluminium have been described, and the reader may have difficulty in deciding what paint or powder-coating system is required. The finishing of aluminium for the building market is a large market and one which has expanded rapidly due to improvements in coatings and the changes in application techniques. Various factors have to be taken into account, and they may be looked at differently, depending on whether one is a metal supplier, an applicator or a user. This chapter will attempt to cover the advantages and disadvantages of each process, the good points and the bad points, and a summary of these are given in Table 11.1. However, as equipment and products improve, manufacturers are able to develop and improve their techniques and narrow the gap.

As the largest sector of the market is for building, this aspect of organic coating of aluminium is the field which will be considered, although particular situations may have exceptional features.

11.2 STOVED WET-PAINT SYSTEMS

This method is the original, and considerable improvements have been made in methods of application such as airless spraying and electrostatic coating to improve the uniformity of paint coating.

TABLE 11.1. *Selection of Paint*

Type of Paint	Advantages	Disadvantages
Air-Drying Liquid Paint	Special coatings for particular circumstances available Applicable on-site Flexible processing operations Large colour range Low cost No size limitation Suitable for small batches Large range of paint types and finishes	Less durable Higher labour costs Several coats necessary Longer drying times between coating
Stoved Liquid Paint	Durable and suitable for architectural use PVF_2 paints available for maximum durability More controlled application Harder and more abrasion resistant coatings Large colour range Medium to low cost of application Suitable for coil-coating of large sheet batches	Medium cost of capital equipment Medium to high production rates Energy required for stoving
Electropaint	Consistent coating thickness Improved coverage at sharp edges Low fume emission on stoving High production rates Low application cost No effluent problems from pretreatment Good "throwing-power" Simple pretreatment before painting	High capital cost of equipment Restricted colour range Technical control of application necessary Assured production rates desirable
Stoved Powder	Thick and durable coatings, suitable for architectural use Range of colours available Low fume emission on stoving Capable of automation Suitable for extrusions in particular	Medium cost of capital equipment Medium production rates Thinning of coat at sharp edges Colour changeovers longer than for liquid paint

Advantages

1. Capital costs of equipment for a hand-operated plant are low.
2. The process is flexible, allowing change of paint system or colour to be made with relative ease.
3. Repair of processing faults is easily carried out.
4. A high quality finish can be achieved.
5. Painting costs are low.

Disadvantages

1. Use of volatile solvents can cause atmospheric pollution on stoving.
2. The finish on a hand plant is dependent on the skill of the operator.
3. Thick coatings can only be achieved with multiple coats.
4. Costs of solvents tend to increase and can cause pollution problems.
5. Poor resistance to handling damage.

With improved paint formulations such as polyester and PVF_2 based resins, more durable coatings are now available as well as improvements made in the formulation of longer established paints based on alkyds and acrylics.

For large areas of sheet, coating of individual sheet by electrostatic spraying or coil-coating by a roller coater is still the most economical method of applying paint, and using the correct procedures and a suitable paint, extremely long life finish is possible, due to a combination of corrosion resistance of the aluminium and the durability of the paint.

A choice sometimes has to be made when using profiled building sheets whether to have the product fabricated from coil-coated sheet, or to have the profiles (produced either by roll-forming or press-braking) manufactured first and have the paint applied subsequently. For ease of production, the first option has advantages, but even using flexible paints, stresses in the paint arise at the formed areas which could lead to local break-down of the paint. The more expensive PVF_2 formulations of paint are more flexible, have good chalking resistance, and these have reduced the risk.

On the other hand, post-painted sheet, i.e. profiled sheet on which the paint is applied after forming, does not have areas of stresses in the paint, the paint does not have to be flexible, therefore, the paint does not have to be a compromise and can be chosen solely for its durability.

11.3 ELECTROCOATING

Its main feature is that it is a method capable of producing a large output with minimum use of labour. It is primarily suited for treating extrusions since, like anodising, maximum treatment efficiency can only be achieved when the full capacity of the plant is taken up. Extrusions have a high surface area per unit of space they take up, which means that the paint bath can be loaded up to its full capacity. With sheet it is not possible to load the bath and make use of all its painting potential.

Advantages

1. Uniform coverage of paint over all areas, particularly on edges.
2. No effluent problems from chromate containing pretreatment chemicals.
3. Only etch cleaning required.

4. Extremely good adhesion achieved between paint and aluminium.
5. Low labour costs.
6. High output possible.
7. Suitable for automatic control.
8. Closely controlled coating thickness.
9. Good defect-free finish of paint possible.
10. Very efficient paint usage.

Disadvantages

1. Limited maximum coating thickness of 40 μm.
2. High capital costs of plant.
3. Close control required for paint with regard to pH, composition and contaminants.
4. Restricted to one colour per paint bath.
5. Maintenance of ancillary equipment such as de-ionisers and ultrafiltration units.
6. Acrylic paints more liable to chalking, change in gloss and loss of colour on weathering.
7. Due to thinness of coating and uniformity of deposition, any metal blemishes such as damage or die-lines originally present are reproduced on the paint surface.
8. Poor resistance to handling damage.

Electropainting is an ideal method of pretreating aluminium in long lengths intended for subsequent fabrication. Consequently, this method is the one favoured by the aluminium extruder who perhaps already has an anodising plant and can electropaint for a similar market by using a process which has features similar to that of an anodising plant.

Attempts are being made to enable thicker coatings of around 50 μm to provide better resistance to handling damage, an advantage at present being enjoyed by powder coating.

11.4 POWDER COATING

Coating with electrostatic powders for architectural applications has been used in Britain since the mid-1970s, and earlier on the Continent. Its use on buildings has been due largely to the development of polyester resins which provide a durable and tough finish, but also to the impetus given by the increasingly high cost of solvents in other paint systems.

Advantages

1. Weather-resistant polyester coating.
2. Thick coatings, 50–100 μm.

3. Because of the thickness, extremely good resistance to mechanical damage.
4. Wide availability of colour.
5. Good obliteration of metal surface defects.
6. High efficiency of powder usage.
7. No pollution from solvents.
8. Suitable for small hand-operated plants, as well as large automated lines.

Disadvantages

1. Inability to obtain uniform coating thickness.
2. Tendency to produce "orange-peel" finish.
3. Thickness of coating tends to restrict width of narrow openings.
4. Effective chromate pretreatment necessary for good adhesion.
5. High cost of powder.
6. Thinner coatings obtained on sharp edges.

Powder coating, when applied correctly and using a satisfactory polyester resin based powder, can provide the most durable organic coatings for finishing extrusions. The thickness of the coating, although a disadvantage from the cost angle, is of considerable benefit in providing resistance to mechanical damage. This is of large importance when coated extrusions have to be transported before final machining, cutting and assembly operations, followed by erection. In order to reduce costs, attempts are being made to reduce the coating thickness below the minimum of 50 μm at present. Another problem is to ensure uniformity of coating thickness over the whole area of an extrusion which at present can vary from 50 to 100 μm.

A comparison between anodising and organic coating can be made, but in many ways the two processes are complementary rather than competitive.

A common factor is that they both provide long lasting and decorative finishes. Anodising gives a harder coating and can provide a more durable finish, but the colours are restricted to natural finish, gold, bronze and black, with a few other colours produced by certain organic dyes. By the nature of the process, light is reflected in a different way and a perfect match between individual components cannot be achieved, since the colour as seen by the eye depends on the viewing angle.

Organic coatings enable one to achieve a very good colour match between individual components, but even the best paints change colour on exposure to weathering and the onset of chalking. The change in colour through chalking should be slight and uniform and it should maintain a homogeneous appearance. Some paints are more resistant to chalking than others, and the effect is more noticeable on dark colours than on pastel shades. The advantage of white paint is that, apart from loss of the initial gloss level, chalking is less noticeable. It would be possible to restore the original colour by lightly abrading the surface, as used for restoring faded paint work on cars with abrasive compounds, but this would be quite impracticable on a building.

The polyvinylidene fluoride resins (PVF_2) used for certain high quality paints have been shown to be very resistant to chalking, and maintain appearance for a long time before the onset of chalking. As mentioned previously, they are expensive and are more difficult to apply.

Laminated coatings of vinyl or PVC do not perform as well as paint coatings on outdoor exposure, the main problem being that they depend on the adhesive used for bonding, and problems can arise over failure or blistering at edges of the sheet. Being thick, they are resistant to mechanical damage, but the plastic film tends to harden and embrittle on exposure to sunlight, and the film can become more easily contaminated by dirt deposits.

11.5 MAINTENANCE OF ORGANIC FINISHES

Protective tapes based on paper or vinyls with low tack adhesives have been used for painted aluminium; these can be stripped off after erection on site. The difficulty in using them is that the adhesive can harden under sunlight, and not only may be difficult to remove, but can strip off the coating if adhesion between the coating and aluminium was poor. In order to avoid this, a layer of wax is sometimes applied, but because of the difficulties, tapes are not often used for protecting painted surfaces.

All architectural finishes including painted aluminium benefit from the washing action of rain. Nevertheless, where dirt and grime become coated on the paint, a soft brush and water can be used to restore the surface. A neutral detergent can be added to the water to assist, which should be cleaned off with fresh water. Abrasives should never be used, nor, in general, should solvents. If the paint is a stoved version, judicial use of white spirit or methylated spirits on a clean rag can be used to wipe off grease or oil, but these should be used with care, and only after testing a small area first.

11.6 OVER-PAINTING OF ORGANIC COATINGS

In certain circumstances, it may be necessary to re-coat on-site aluminium which has previously been painted, such as for a change of colour.

If the original paint is of good condition, all that is necessary is to wash down, and dry before applying one or two coats of good quality air-drying paint. If possible, it helps if the original surface is lightly abraded to form a key for the subsequent paint. The air-drying finish will not of course behave in the same way as the underlying stoved finish and will degrade at a faster rate.

11.7 REPAIR OF DAMAGED AREAS

Organic-coated aluminium, whether treated by a stoved paint finish or powder coating, may sometimes suffer from damage which could have occurred during transport to site, during fabrication or during erection. This might result in small areas where the coating has been removed and exposing the underlying aluminium. Removing the fabrication and re-coating and re-stoving the component would be disproportionally expensive, and an acceptable method of repairing the damage on-site is to be preferred wherever possible.

It should be realised that the rectification carried out will not provide a surface with the same weathering characteristics as the original stoved surface. An air-drying paint may be a perfect colour match when applied, but weathering may change the appearance more rapidly than the original surface, such as loss of gloss, greater presence of chalking, and change in colour. Nevertheless, as the damaged areas will be small, the slight differences will be usually acceptable.

11.7.1 Repair Procedure

1. If necessary, feather the edges of the damaged area with emery paper ("wet and dry")
2. Degrease with white spirit by wiping and dry
3. Apply a coat of 2-pack etch-primer and allow to dry
4. Apply a coat of zinc chromate metal primer
5. Apply a finish coat

The finish coat should be a good quality air-drying external paint. Amongst the more durable of these is a 2-pack polyurethane gloss finish.

These procedures are applicable to powder coatings as well as stoved alkyds, acrylics and polyester paint finishes.

Touching-up of PVF_2 coatings is also similar, but the problem is that owing to the pigments used for this type of paint, the colours obtained do not always conform to those given in BS 4800 which most architectural paint finishes follow. In this case, it is often possible to use one of those from the range offered for car finishing which can provide a better match.

Chapter 12

Testing of Organic Coatings on Aluminium

12.1 INTRODUCTION

The quality of aluminium coated with a paint or a fused powder coating depends largely on two factors, the coating itself, and its application to the aluminium. Both factors are critical in obtaining a satisfactory product.

The paint and its formulation are in the hands of the manufacturer, but it is the responsibility of the applicator to ensure that the quality of the finished product is up to standard. This can only be achieved by careful sampling and testing of the raw material before it is applied to the aluminium. If this is done, it will ensure that the specification has been met, and will avoid the costly application of an incorrect coating to aluminium.

A number of tests are available to test the quality of coatings. These should be carried out with two objects in view. The first is to ensure that the paint or powder is itself satisfactory, and the second is to ensure that the paint applied to the aluminium has been processed under satisfactory conditions.

This distinction between testing of the raw material and testing of the coated product is recognised in BS 4842 and BS 6496.

Testing of all incoming coating materials may seem an expensive or unnecessary overhead. Some hold the view that the materials are the responsibility of the supplier, who himself is likely to be carrying out thorough checks on his product. But problems can still sometimes arise, and in the case of a resultant dispute, the supplier may argue that he is only liable for the value of his product and not for any consequential loss. The latter can be disproportionally high when taking into account the cost of replacing, reprocessing and perhaps, in the case of architectural components, re-erection.

In practice, a compromise is often reached when the material supplier supplies a test certificate with each batch, giving the results of tests he has carried out. The purchaser therefore need only occasionally carry out tests on the raw materials as a spot-check or where there might be some doubt on the results given in the test certificates. This procedure is very effective when good relationship exists between customer and supplier.

Standards exist for the testing of paints, and these, obviously, are not confined to paints used for aluminium. BS 3900: "Methods of Test for Paints", is a comprehensive document describing tests on both liquid paint and on the physical properties of the cured paint coating itself, enabling one to test a large variety of properties.

In this specification, the tests are divided into seven groups as follows:

Group A — Tests on liquid paints (excluding chemical tests)
Group B — Chemical examination of liquid and dried paint film
Group C — Tests associated with paint film formation
Group D — Optical tests on paint film
Group E — Mechanical tests on paint film
Group F — Durability tests
Group G — Environmental tests.

Many of the tests are based on DEF 1053, "Standard Methods of Testing Paints, Varnish & Lacquers and Related Products" required by the Ministry of Defence.

It is the last four groups that are of more interest to the specifier or user of painted aluminium products.

12.1.1 Group D — Optical Tests

D1 The first part describes the conditions which the test pieces must undergo before examination, as colour can change under different environments.

D2 Gloss (specular reflection value). The method of determining gloss given is identical with the International Standard ISO 2813. It describes instruments which can measure gloss, using instruments with angles of incidence of 85°, 60° and 45°, depending whether the paint coating is of low gloss or high gloss.

An architectural paint finish will be of relatively low gloss, at about 30°, although the paint coatings on extrusions used for window sections may be higher.

On a production line, it is usual to compare visually the coating under test with a sample of known gloss.

12.1.2 Group E

The mechanical tests described in BS 3900 are:

E1 Bend test
E2 Scratch resistance test

E3 Impact (falling weight) test
E4 Cupping test
E6 Cross-cut test
E10 Pull-off test for adhesion

Most of these determine the physical properties of the paint coating. The scratch resistance test is of importance as a measure of wear resistance and E1, E3 and E4 are a measure of formability, more of interest where the coated aluminium is to suffer deformation after coating, such as may be carried out on coil-coated aluminium. The last two determine the adhesion of paint to aluminium.

The cross-cut test is one of the most useful and easily carried out tests. It is a measure of the adhesion of the coating to the aluminium, and can be carried out both on-site or as a test during production.

The test consists of scribing a grid on to the paint surface, applying an adhesive tape and pulling away sharply. If the adhesion is good, no paint removal results.

The standard describes the cuts to be made, which are in the form of a grid with either 6 or 11 cuts in both directions to form a grid of 25 or 100 squares, the cuts being either 1 or 2 millimetres apart. The method of making the cuts, applying and removing the tape, is given.

A chart is given showing 6 illustrations with increasing amount of paint removal which allows one to classify the result of one's test, ranging from 0 with no paint removal to 5 with more than 65% of the squares being affected.

This test is technically identical with ISO 2409. The Pull-off Test for adhesion (identical with ISO 4624) describes how adhesion can be measured, by using an adhesive to stick a rod of steel to the paint and physically pulling off the paint on a tensile testing machine.

12.1.3 Group F

Some of the tests here are concerned with coated aluminium. These are:

F2 Resistance to humidity under condensation conditions
F3 Resistance to artificial weathering (enclosed carbon arc)
F4 Resistance to continuous salt spray
F6 Natural weathering tests
F8 Resistance to sulphur dioxide
F9 Resistance to continuous condensation
F10 Resistance to liquids

The tests determine the resistance of paint to various environments.

The first humidity test depends on exposing painted samples at 100% humidity and 42° to 48°C every hour followed by cooling to ensure intermittent condensation for periods up to 10 days. It is of less importance to painted aluminium than to painted steel.

Artificial weathering test ("Weatherometer") is a rotating drum on which the samples are exposed, at the centre of which is a carbon arc lamp. A fine water mist spray is introduced, and the temperature is controlled. The duration of the test is about 3 months, and it will give an indication of how a paint will perform exposed to the weather, and whether or not chalking of the paint is liable.

Warning is given that this test gives results that show that not all paint types will correlate with actual conditions of exposure, but used intelligently, it gives useful information.

The continuous salt spray test is based on DEF 1053 Method 24, Resistance to Continuous Salt Spray. It describes a continuous salt spray cabinet in which painted samples are subjected for 10 days or more. The test is known as the Neutral Salt Spray Test (NSS).

The salt solution consists mainly of sodium chloride (26.5 g/l), but this solution is rather mild for testing painted aluminium. More meaningful results for aluminium are given using the above solution with additions of acetic acid (ASS). The two tests are described in BS 5466 pts 1 & 2 (Methods for Corrosion Testing of Metallic Coatings) and described later under 12.3.2.

Notes are given in BS 3900 (F6) on how to carry out natural exposure tests of samples, such as how they should be mounted, and how results should be interpreted. It also gives useful information on exposure sites and on how to determine the level of atmospheric pollution.

Resistance to sulphur dioxide is an important factor when considering paint for industrial atmospheres. The test described is technically identical with ISO 3231, and consists of exposing test pieces to an atmosphere containing an agreed amount of sulphur dioxide at 40°C for 8 hours, followed by 16 hour rest period. The cycles are repeated for a specified number of times, the samples being examined every cycle.

The second humidity resistance test is identical to ISO 6270 and depends on continuously exposing samples to an atmosphere maintained at 35°C over a bath of water at 40°C. The test panels form a window, the painted surface on the inside, and the reverse surface exposed to the outside so that there is continuous condensation on the paint.

12.1.4 Group G

These environmental tests are more particular to the use to which the painted surface might have to undergo.

G5 Resistance to liquids
G7 Determination of the effect of heat

Applicable to aluminium, the test determining the resistance to liquids, technically identical with ISO 2812, describes various methods that can be used to determine the resistance of a painted surface to a liquid, either by immersion in the liquid, or by using an absorbent material saturated with the liquid.

For determination of the effect of heat, the test method, technically identical to ISO 3248, describes how a sample is heated in an oven at 125°C for 24 hours and then examined. It also allows for mechanical testing of the paint to be carried out after heat-testing when required.

A number of specifications have been issued to govern the standards to which coated aluminium should conform. Usually, the specification gives the tests that are to be carried out so that the product can be checked, and the level of the test results which must be met.

The first British Standard covering stoved organic coatings specific to aluminium was BS 4842 "Stoving Organic Finishes on Aluminium Extrusions and Preformed Sections for External Architectural Application", published in 1972, and later revised.

It is divided into 2 parts, the first laying down the properties required of the coating applied. The requirements include:

consistency	scratch resistance
stoving time	curing
finish	effect of mortar
colour	resistance to acetic acid/salt spray
gloss	resistance to sulphur dioxide
adhesion	artificial weathering.

The coating is examined by applying the test to a standard panel, under controlled conditions, and carrying out tests on the above properties.

The second part covers the requirements for the finished product:

film thickness	adhesion
finish	curing
colour	resistance to sulphur dioxide
gloss	

BS 4842 when originally published was intended to cover aluminium that had been coated with a liquid paint, either by spraying, roller-coating, or by electrophoresis.

The main changes in the revised version of the specification are to replace some of the tests with those taking a shorter time, and an important step is to stipulate the pretreatment to be applied before coating, this being a chromate conversion coating, except in the case of electrodeposited coatings.

The tests required by this specification are based on those given in BS 3900 where applicable, with the following additions, given as appendices:

12.1.5 Mortar Test

This consists of applying a pat of wet mortar to the painted surface, made up as follows:

37.5 g lime
37.5 g cement
225 g sand

The ingredients are mixed together and water is added to make a paste. Pats of mortar, about 12 millimetres in diameter, are applied to the painted surface and stood for 24 hours at 38°C and 100°C humidity.

FIGURE 12.1. Portable digital eddy current meter for measurement of non-conductive coatings on aluminium — Eleometer Ltd

The parts are dislodged and the painted surface is examined for discoloration or loss of adhesion, of which there should be none.

12.1.6 Salt Spray Test

The neutral salt spray test given in BS 3900 is considered too mild for painted aluminium, so an acetic acid salt spray is used, consisting of the following solution:

5 parts by mass of sodium chloride
95 parts by mass water
Add 6 ml of glacial acetic acid per litre of the above solutions.

The test is carried out as described in Appendix E of BS 4842. There shall be no blistering, softening or loss of adhesion after exposure for 24 hours, and there shall be no more than 1.5 mm of corrosion creep from a previously scored line.

12.1.7 Curing

The assessment of curing is determined by rubbing the painted aluminium firmly with cotton wool soaked in methylisobutylketone. After 30 rubs there shall be no more than a slight marring of the finish.

At the same time, opportunity was taken to cater for powder coated aluminium, as there are certain differences, largely due to the increased thickness of the coatings on aluminium when treated this way. This aspect is now covered by BS 6496.

One difference with regard to powder coating was the realisation that the coating thicknesses resulting from this method can be between about 50–100 μm.

Another important introduction was the Permeability and Quality of Pretreatment Test. This requires the powder coated sample to be partially immersed in water in a pressure cooker at 100 kPA (= 15 lb/in^2) for 2 hours where no blistering of the coating shall occur except with 3 mm of the edge of the sample panel. This test has found use in checking the quality of the adhesion of powder coatings.

12.2 SPECIFICATION FOR STOVED ORGANIC FINISHES ON ALUMINIUM WINDOW PRODUCTS

This is a specification published by the Aluminium Window Association, intended to complement BS 4842: 1972, and was published in 1976. It defines the responsibilities of the coating supplier and the applicator.

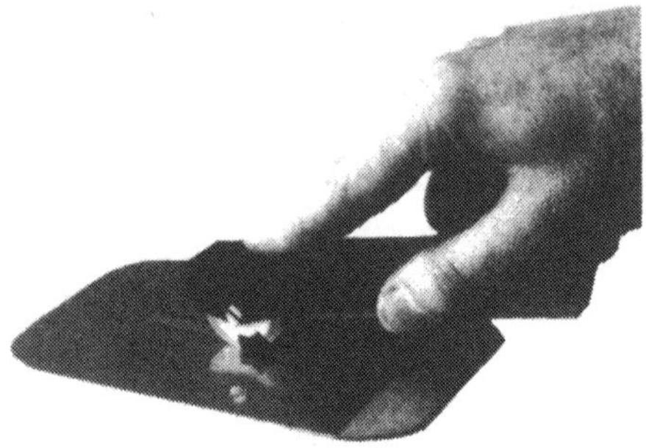

FIGURE 12.2. Cutter for cross-hatch test — Sheen Instruments Ltd

It requires the coating supplier to provide a certificate that the materials to be applied have been tested and conform to the requirements of BS 4842. It also requires a further test to be carried out on the resistance to detergent, and this test is given in an appendix.

The appendix describes a detergent solution, in which the coated aluminium is immersed for 72 hours, followed by application and removal of an adhesive tape. No loss of adhesion or change in appearance should occur.

This specification generally follows the requirements of BS 4842 and with similar tests. It lays down the need for a chromate conversion coating except where electrophoretic coating methods are used.

Emulsion cleaning or trichloroethylene cleaning prior to conversion coating is permitted, but in practice chemical cleaning is necessary for satisfactory results.

The tests to be carried out on the coated products are as follows:

1. Appearance.
2. Colour and gloss. Comparison is made usually with an agreed sample.
3. Thickness. Measured by an eddy-current meter, the minimum is 20 μm except for PVC Organasol or Plastisol Coatings which shall be at least 100 μm thick.
4. Adhesion. The cross-hatch and adhesive tape pull-off method is used.
5. Hardness. Either by pencil hardness to resist an H pencil or by the Scratch Test of BS 3900:Part E2.
6. Curing. Rubbing for 30 cycles with cotton wool and methylethylketone (MEK).
7. Water resistance. This requires a test piece to be partially immersed in water in a pressure cooker and heated to boiling at a pressure of 0.10 N/mm^2 (15 lb/in^2) for 2 hours. The sample is removed and cooled, examined, and an adhesive tape is applied and removed after 1 minute.

No coating removal, or blistering other than within 3 millimetres of the edge, should occur. This test is of particular use for powder coatings to give an indication of the adhesion.

8. Sulphur dioxide. This is slightly different to that in BS 3900 and is described in BS 1391:1952. It consists of exposing samples in a heater at a controlled temperature in an atmosphere of sulphur dioxide for 24 hours or more.
9. Mastics and sealants. Suitability of these products are the responsibility of the user.

The specification gives advice on maintenance of coated windows. It also gives advice on handling window frames before and during erection.

It advises strongly against the use of lacquers or protective tapes for temporary protection because of the difficulty of removal with possible damage to the organic coating.

12.3 PERFORMANCE REQUIREMENTS AND TEST PROCEDURES FOR PIGMENTED ORGANIC COATINGS ON EXTRUDED ALUMINIUM

This specification was published by the Architectural Aluminum Manufacturers Association of America in 1972. The tests and requirements are similar to BS 3842, and uses tests published by The American Society for Testing Materials (ASTM).

In brief the tests and requirements are:

1. Colour — Visual comparison
2. Gloss — Using a 60°C gloss meter
3. Hardness — H pencil minimum.
4. Adhesion — Cross-hatch and tape removal.
5. Impact resistance — An impact load, using a Gardner Variable Impact Tester, is applied to the painted surface to deform the sample by 0.10 inch, followed by application and removal of tape over deformed area. No removal of coating is permitted.
6. Acid resistance — 10 drops of 10% (vol) hydrochloric acid is applied to the painted surface for 15 minutes, and no change in appearance or loss of adhesion should occur.
7. Mortar test — similar to that of the Aluminium Window Association, and based on ASTM C207.
8. Detergent test. The composition of the solution is different from the Aluminium Window Association test, but the principle is the same.
9. Humidity test. Samples are exposed continuously in a 100%

humidity cabinet at 100°F for 1000 hours (US Federal Test Method No. 141a, method 6201). Only a few blisters are permitted.

10. Neutral salt spray. Samples previously scored are exposed for 1000 hours, and an adhesive tape is applied. No failure of paint should occur when the tape is removed, other than at a distance of not more than 1/10 inch from the cut.
11. Accelerated exposure. Using a Weatherometer, samples should withstand 500 hours with no loss of adhesion and only slight colour change or chalking.
12. Outdoor exposure. Samples are required to withstand one year's exposure in fluoride with only slight chalking or fading and no checking, crazing or loss of adhesion.

The standards described offer a complete system of tests that can be carried out, and they provide the levels to which the coated aluminium must reach. The tests are a compromise between the long-term outdoor testing at different types of exposure sites to determine the expected life of a product, and the rapid results required by a coating applicator who requires to know how his product will perform when produced under the conditions he applies.

Short term testing can only give a guidance on ultimate performance; nevertheless, it is an essential part of quality control.

The Pull-off Test of BS 3900, Part E 10, is only for use in a laboratory, and it gives only limited information.

One test developed by British Aluminium Co. (now British Alcan Aluminium plc) and used as control for their Colourbond painted sheet, consists of immersing samples in phosphoric acid, under the following conditions:

Phosphoric acid	10%
Temperature	25°C
Time	28 days

Coated aluminium is immersed in the solution, having previously had a score cut in the painted surface. An adequately stoved paint will withstand 28 days in the solution with only some deepening of the scores and no undercutting of the paint adjacent to the scores.

No lifting or blistering of the paint occurs on areas away from the scores. Any uncoated areas of aluminium must first be protected before testing, otherwise the strength of the acid is reduced.

This test has been used for 20 years and has proved very effective on determining the adhesion between a stoved finish and the aluminium, not only on alkyd and acrylic finishes, but on epoxies and polyester powder coatings. It will show up failings in pretreatment as well as paint application. In some instances of poor adhesion, defects will arise after 3–4 days immersion.

The test is too severe for use with most air-drying systems and possibly with some coatings which are not acid resistant.

The tests devised are a result of experience gained in the manufacture of coatings and in their application.

The amount of testing that should be carried out must depend on a number of factors. Ideally, all the tests should be carried out on every production batch, but this would be impractical where some of the longer term tests are concerned. Even the comparatively short time of 72 hours is far too long for most production purposes, such as for a jobbing applicator. Nevertheless, even he should carry out the full range of tests occasionally to ensure that his production processes are up to standard.

Certain tests should certainly be carried out frequently for production control purposes: they are fast, give immediate results, and measure the more important features which are under the control of the processor.

Coating thickness	By eddy current meter
Hardness	By pencil hardness test
Adhesion	Cross-hatch and tape test
Solvent resistance	Rub test with cotton wool and MEK (or acetone)
Gloss	Comparison with a standard
Colour	Comparison with a standard

Hardness, solvent resistance and gloss results all depend on stoving conditions. Under-curing will be indicated by a soft coating and poor solvent resistance. Over-curing will be indicated by change in colour and a very hard result on pencil test.

Adhesion is one of the most important properties for an organic coating on aluminium, and it is not easy to obtain accurate information. The cross-hatch test is rapid and easy to carry out, but adhesion has to be poor for failure to become apparent.

Adhesion properties of powder-coated aluminium are most important, which is very much dependent on satisfactory pre-treatment and on stoving.

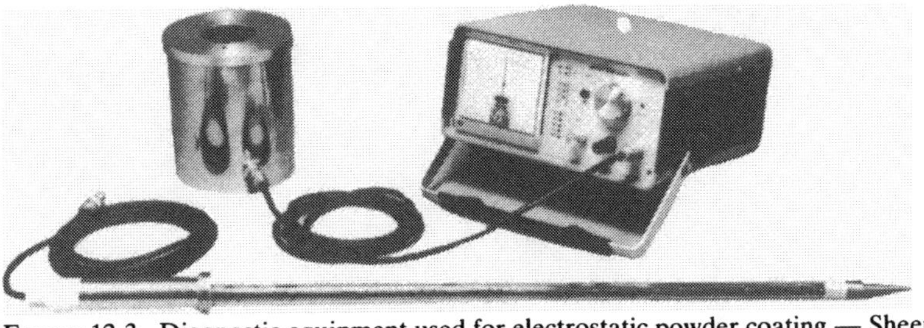

FIGURE 12.3. Diagnostic equipment used for electrostatic powder coating — Sheen Instruments Ltd

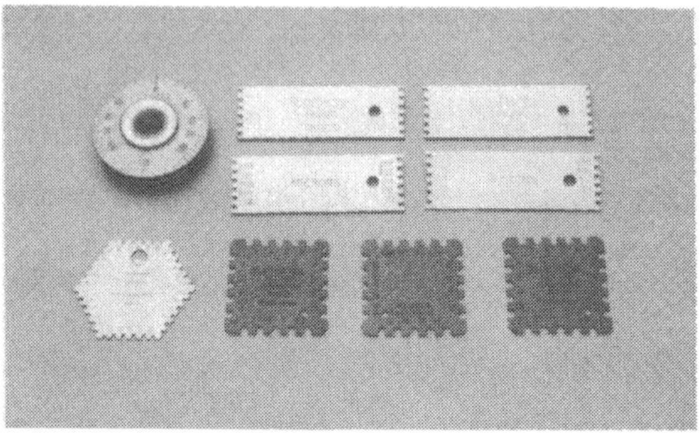

FIGURE 12.4. Gauges used for measurement of wet paint films before drying or stoving — Eleometer Ltd

The pressure cooker test of BS 4842 is one which can be carried out without too much difficulty, and it gives a useful indication on the quality of the product.

Another test for powder-coated aluminium, used more on the Continent, is known as the Machu Test. One version of it consists of applying a cross-hatch to the coated sample and immersing in the following solution:

Sodium chloride	60 g
Hydrogen peroxide (5%)	5 g
Acetic acid (glacial)	10 ml
Water	1 l
pH	3-3.3
Temperature	room

When immersed in the above solution at room temperature for 24 hours, no creep or loss of adhesion should occur at cut edges.

Chapter 13

Coatings for Aluminium used in Packaging

13.1 INTRODUCTION

One of the largest markets for rolled aluminium products in the UK is packaging, both for food and other articles. About 50,000 tonnes of aluminium are used as foil, and 68,000 tonnes for other packaging components such as cans and can ends. Both of these outlets combined make up over 40% of the total requirements of rolled aluminium. Taking into account the

FIGURE 13.1. Decorated aluminium beverage cans. The bodies were drawn from strip followed by ironing of the walls

thin gauges used for both foil and containers, these figures represent a very large surface area, hence the great importance of surface finishing where many of the products undergo a highly decorative surface treatment process.

Aluminium cans have been used for some years; initially the largest increase occurred in the United States. In the United Kingdom the use of aluminium for this market was slower in arriving because of the competitive nature of tin-plate, the original raw material, which was produced here at extremely low prices and put aluminium at a relative cost disadvantage. Due to the increased cost of tin-plate, rising at a higher rate than for aluminium, and improved manufacturing processes, the preference for aluminium became more apparent, and the aluminium can became accepted here.

Surface treatment in this field can be subdivided into two areas, where coatings are applied for

(a) External decoration and labelling
(b) Protection, especially internal.

Aluminium cans are a prime example of packaging; lacquer coatings are applied internally to protect the aluminium from attack, to avoid contamination or to avoid marking or blackening of the contents from contact with the metal. Decorative coatings are applied externally.

13.1.1 Can Bodies

The ductility of aluminium is especially suited to the production of two-piece cans where the body is produced from coiled strip by deep drawing often followed by ironing of the walls to produce thin gauge cans, reducing the weight of aluminium being used and hence improving its price competitiveness with steel. Another advantage of the drawn can body, be it in aluminium or other material, is the absence of a side seam. This avoids the need for soldered or mechanical side seam joints and very greatly reduces the risk of double-seam leakage.

For the food and drink industry the internal surface of the formed cans are protected by a coating of lacquer to avoid contamination of the food from possible dissolution of the aluminium and also to protect the aluminium from pitting or corrosion by the contents. Some foods such as fruit or those containing vinegar are slightly acidic and prolonged storage would result in failure of the aluminium by pitting or pin-holding if it was unprotected.

13.1.1.1 *Can bodies produced from untreated aluminium*

As indicated above, the use of plain aluminium coil for the production of drawn and wall-ironed bodies has significantly increased in recent years.

It is principally used for cans containing carbonated beverages such as beer and soft drinks. The original coil thickness is required in the profiled can base to resist deformation from the quite high internal pressures. The thin ironed can walls are in turn supported by the internal can pressures, thus resisting handling abuse, etc. Unlike other materials such as tinplate, aluminium is essentially a single homogeneous metal. When coil is produced in some of the latest and most sophisticated rolling mills in the world, an exceptionally "clean" material free from inclusions and pinholes is obtained which is also of very uniform thickness both within and between coils. Drawn and wall-ironing lines need this quality of product if they are to run consistently and economically at high speed. To facilitate fabrication the plain coil has to be lubricated before drawing. The resultant cups are then immersed in a lubricant cum coolant during the wall-ironing stage.

In consequence, elaborate continuous washing, drying, surface treatment and lacquering or coating procedures are essential. All traces of lubricant must be removed before commencing the various pretreatment and coating stages, and rinsing operations prior to coating must be thorough to remove all traces of residues from the previous stages. Thus the final rinse before coating must be with de-ionised water to avoid calcium deposits from hard water.

Although the finishing of the internal can surface, usually by lacquer spraying, is a complex operation, it offers a continuous and practically perfect film especially when two separate coats are successively applied.

13.1.1.2 *Can bodies produced from lacquered and/or coated aluminium*

Aluminium in this state may be presented to the can production line as coil or cut sheets. It is generally used for shallow drawn cans or ends (discussed later).

The coatings are applied to aluminium on a coil-coating line which on a modern plant operates at high speeds. If required by the customer as sheet, the coil is slit to width, levelled, and guillotined to length.

With improvements in lacquers, increasing depths of draw can be obtained without crazing or damaging the coatings. This increases the necessity to provide good adhesion between the aluminium and the coatings, and one plant in Germany has developed a method of ac anodising as a pretreatment rather than applying a conversion coating. This process provides better adhesion than the non-chromated conversion coatings now required for the food industry.

Both sides of the coil can be coated, one with a protective lacquer, the other with decorative coatings.

The type of can or container made from precoated aluminium is largely used for containing food, especially for fish, paté, dessert packs, etc.

FIGURE 13.2. Beverage can with "easy open" top

13.2 CAN ENDS

When produced from aluminium, ends are invariably of the easy-opening type — ring-pull apertures for beer and soft drink cans — full apertures for food cans.

Ends may be produced either from sheet which is generally lacquered by the can manufacturer or from lacquered coil prepared by the aluminium manufacturer or by a toll coater.

13.2.1 Ends for Beverage Cans

The ring-pull aperture end is typically produced from 5182 alloy containing 4½% Mg and .3% Mn, and rolled to a hard temper. The tab stock is to a different specification, viz. 5.42 with 3.5% Mg and .3% Mn. For beer can ends an internal lacquer, e.g. epoxyphenolic, may be used. Some metal exposure in the aperture score area may be acceptable and even beneficial where the product is concerned. Metal exposure must be avoided on the finished soft drink can end. Again, it is necessary to pretreat the aluminium surface to ensure adequate lacquer adhesion. A relatively thick coating, of the organosol type, is often used. Alternatively, the internal end lacquer system may be sprayed after scoring.

13.2.2 Ends for Food Cans

These full-aperture ends are made from a different alloy, e.g. 5052 containing 2.5% Mg and .3% Cr, rolled to a hard temper. Process resistant lacquers, etc. are also required.

13.3 GENERAL CONTAINERS

A large number of containers are used for commercial products which have no reaction with aluminium. The only coatings required for these applications are for external identification, marking or decoration. The products in this range do not require the same standards for coating, and many of the pretreatment stages can be omitted. Contents of these containers may be liquid paint, many powdered or solid products and many organic solvents.

13.3.1 Closures

Aluminium alloys are used for production of closures for bottles both as crown caps and screw type. The latter are drawn components which are placed over filled bottles and threaded by rolling the aluminium using the glass bottle as a former. These can be incorporated as a tamper-proof closure.

13.3.2 Finishing

Designs are produced on one side by printing methods, making allowance for the deformation to be finally produced on the aluminium. The inner surface must be protected with a lacquer, and a gasket for sealing can be cut from a vinyl type sheet. Threaded closures often use a plastic or varnish coated paper disc as a gasket which is inserted in the cap.

13.4 COLLAPSIBLE TUBES

Aluminium collapsible tubes are made from impact extruded slugs of pure aluminium. Printing and decoration of the outside must be carried out after forming, and lacquers are used for internal protection. Special requirements arose over the introduction of fluoride containing toothpastes which might perforate unprotected aluminium tubes on storage.

13.5 BEER KEGS

Aluminium beer kegs have been in extensive use for 25 years in a field where wood was the accepted material. Castings have been used, but now

FIGURE 13.3. Aluminium beer kegs internally protected by stoved lacquer using anodising as a pretreatment — Grundy Ltd

fabricated casks are made from 6082 alloy pressings, welded together and reinforced with 6082 extrusions.

Internal protection is required, not so much from the action of some beers but for the cleaning operations between filling cycles.

Anodising, with a coating thickness of 5 μm, is sometimes used, but lacquering with a stoved epoxy lacquer also gives good service. As with all organic coatings on aluminium, adhesion is a most important feature. One method is to apply a chromate conversion coating, the chromium being in the tri-valent state rather than the toxic hexa-valent state which would not be acceptable in the food industry.

FIGURE 13.4. Aluminium for packaging. Foil labels with printing. Also aluminium caps pressed out from preprinted and decorated foil — British Alcan Aluminium plc

An alternative method is to apply a thin anodic coating of ¼–½ μm thickness which acts as a good key to the lacquer which is subsequently sprayed on to the internal surface. The unsealed anodic coating must be rinsed with de-ionised water and dried before coating to avoid contamination from residues of untreated water. The coating, which can be clear or pigmented, is finally stoved.

13.6 ALUMINIUM FOIL

Large quantities of aluminium foil are used in packaging on account of its resistance to UV radiation and its impermeability to water vapour. Much of it is coloured and decorated by printing by offset litho processes, or it can be laminated to paper or plastic film.

A variety of printing processes can be used, and a number of coatings are applied depending on the application and resistance required. Although some coatings can be directly applied to the foil surface, with the proviso that the surface is clean and grease free, it is often necessary to prime the aluminium with a synthetic resin before applying subsequent coating.

Chapter 14

Miscellaneous Finishes

14.1 INTRODUCTION

Many finishes can be applied to a variety of surfaces but the coating procedures may have to be modified when treating aluminium. Some of these finishes may be of a more limited or specialised nature but nevertheless are of interest.

14.2 VITREOUS ENAMELLING OF ALUMINIUM

Vitreous enamelling of metals, particularly iron and copper, is an ancient art. It is a coating based on mixed salts in the form of powdered glass, known as frit, that is applied to the metal which is then fused under high temperature to form a glass-like coating, hard and durable, but not very flexible.

Normal enamels applied to most other metals are unsuitable for aluminium as the fusion temperature of the frit is near or above its melting point. Special frits were developed for aluminium which had a lower fusion temperature at around 550°C, but even at these temperatures the aluminium becomes annealed and loses much of its potential strength. Some of the original low-melting enamels used for aluminium contained lead salts which made them unsuitable for cook-ware.

Aluminium foil, 0.3 mm thick, has been coil-coated with a sufficiently flexible vitreous enamel intended as facings for insulating board. Vitreous enamelling has been used as a coating for aluminium domestic components required to withstand heat, such as the casings of toasters.

Vitreous enamelled aluminium has been used for architectural applications as a means of applying a coloured coating to aluminium sheet, sections or castings as cladding on buildings, providing a durable and non-fading surface, which is easily cleaned. Vitreous enamelled cast aluminium pillars produced in 1929 are still said to be in good condition. Vitreous enamelled aluminium has also been used on American warships, and in the 1960s a lighthouse in America was clad with enamelled aluminium.

Enamelled aluminium for architecture has largely been replaced by powder and paint coatings which, although perhaps not so durable, are considerably cheaper and free of some of the problems that can arise over application of enamels.

One field where vitreous enamelling is used to advantage is coating of the outside of both cast and wrought cooking pots which provides excellent resistance to heat and abrasion. Heat resistant polyimide coatings have sometimes been used for cooking pots heated on electric cookers, but these coatings are not suitable for utensils on gas cookers where flame may be directed on to the sides and would harm the paint. Lead-free frits were therefore developed for this application and can provide a range of colours. The coatings can be sufficiently flexible, to allow the rim of a pan to be turned over after coating as a stiffening effect to make the annealed pan more rigid.

Vitreous enamelling is used as an internal lining for pans used in conjunction with PTFE non-stick coatings to provide wear resistance. (This is described later under 14.3.)

14.2.1 Application

The frit is manufactured by fusing a mixture of quartz with soda, potash, boric acid etc., to which titanium dioxide is added as an opacifier.

After fusion and mixing, it is quenched in water and finely ground with inorganic pigments and titanium or lithium oxide to form the frit.

This is mixed with water to form a slurry and is applied to the aluminium by spraying. After drying, the components are fired in an oven at about 500°C to fuse the coating into a dense, glass-like layer.

14.2.2 Affect of Alloy

Pure aluminium, and the manganese containing alloys (3103 and 3003), are suitable, providing the copper content is less than .15%, and the magnesium less than 0.005%. Magnesium in aluminium is a problem for vitreous enamelling as, during firing, it tends to migrate to the surface, forming magnesium oxide, which is not soluble in the fused frit and causes problems of adhesion between the aluminium and the coating. Magnesium silicon alloys (6000 series) have been coated successfully, provided a chromate conversion coating process has been first applied. The pretreated aluminium must first be prefired before applying the frit.

Aluminium magnesium zinc alloys (7000 series) are not easily coated.

As sheet, these alloys have been clad with pure aluminium to provide a suitable grade for enamelling. Alternatively, the unclad alloy can first be treated by a chromate conversion process. These alloys can undergo a certain amount of natural ageing after firing of the enamel which allows them to regain some of their strength.

14.2.3 Testing

A problem of vitreous enamel is through failure in adhesion where a fault known as spalling can arise. This was one of the reasons why development trials in coating aluminium sheet for architecture in the mid 1960s came to an end.

Spalling is the cracking and detachment of enamel which can occur at edges of an enamelled sheet, sometimes also on other areas where mechanical damage has arisen.

A useful corrosion test is to immerse samples in 5% ammonium chloride or 1–2% antimony trichloride for 24–100 hours. If adhesion is satisfactory, little or no spalling should occur at cut edges of the specimen. Other tests given in BS 1344 (Methods of Testing Vitreous Enamel Finishes) have been adapted for use with aluminium.

14.3 ELECTROLESS COATINGS ON ALUMINIUM

Electroless or chemical plating is, as the name implies, a method of depositing a metal from a solution of one of its salts without the need to apply an electric current. It takes place through the catalytic reduction and deposition of nickel from solution on to the surface of aluminium which has first been treated by a zincate process, as for electroplating.

The process was originally developed in the United States where nickel was deposited on to metals from hot acid or alkali solutions based on nickel chloride. It is suitable for treating aluminium to deposit nickel on to the surface. While it tends to be more expensive than electroplating, it has the very important advantage of having extremely good "throwing power". In other words, it is capable of depositing nickel in recessed areas where conventional electroplating would be ineffective due to shielding. It can be used to plate nickel even on the interal surfaces of narrow tubes, provided flow of solution is maintained through the bore.

Most of the nickel coatings deposited by electroless methods contain 5–15% phosphorus and are more corrosion resistant than electrically deposited nickel coatings of equal thickness, aided also by reason of the coatings being more uniformly deposited and of low porosity. Electroless coatings have better properties of lubricity compared with electrodeposited coatings.

The brightness of the coatings depends on the purity of the solutions used; some commercial products are better than others. The coatings, though, are not as bright as those obtained electrolytically.

The deposit of nickel, plated by these processes, is hard and abrasion resistant, but maximum hardness is achieved by heat-treating the plated components at 230°C for 30 minutes, where a hardness of 600 VPN can be achieved.

14.3.1 Applications

The deposited coating of nickel, whilst being hard, can impact corrosion resistance in certain circumstances. Nickel plating has been used for increasing the wear resistance of aluminium piston heads. It has also been used with high speed aluminium pistons for coating the grooves which hold the rings. The hard coating protects the aluminium from wear in service.

Other engineering applications include hydraulic cylinders, moving pump components, lock nuts, shafts, slide valve plates and threaded parts, where not only wear resistance is required, but advantage is taken of the lower frictional resistance. Coating thickness of 12–130 μm may be applied.

In the electrical field, aluminium components may be coated with nickel to allow easy soldering with lead-based solders, where nickel coatings of 7–20 μm may be specified. It also provides good electrical conductivity for joints.

The system has been used for manufacturing small complex parts, where aluminium formers have been used which are electroless nickel plated, followed by dissolving out the aluminium in caustic soda, leaving the shell of nickel.

14.3.2 Processing

Since no electricity is required, components need not be racked, although for larger pieces, and where uniformity of coating is necessary, stainless steel or aluminium jigs are required. Small components can be treated in bulk by barrelling.

Before application of nickel, the aluminium must be treated by a zincate pretreatment in the same way as prior to electroplating, with the same precautions and variations for treating different alloys. This is described in Chapter 4.

The nickel solutions are acidic and should be contained in stainless steel tanks, preferably fitted with removal polythene liners. Before use the stainless steel should be passivated for 2 hours with concentrated nitric acid. Titanium tanks can be used and are slightly superior to stainless steel, but the metal must be anodically protected at a current density of 5 mA/dm^2.

Polypropylene can be used but requires immersion heaters for the solution.

The solutions are operated hot, at about 90°C, requiring evenly applied heat and agitation during operation, either by air or mechanically.

The operating conditions will vary according to the suppliers instructions, but usually the nickel concentration is maintained around 6.5 g/l, and the pH is kept at around 4.5.

Exhaust extraction is desirable to remove steam. Filtration of the solution during operation is preferred, which helps with circulation of the solution and stabilisation of the heat as well as removing solid impurities.

14.3.3 Maintenance

During use, the depleted salts are replaced by additions of further chemicals according to the analysis procedures advised by the supplier. This will usually maintain the pH at the correct level, otherwise it can be rectified by additions of acid or alkali.

The deposition rates depend on the state of the solution; a new bath will deposit 15–20 μm/hour. The surface area of aluminium being treated should be approximately 2 dm^2/l of solution.

14.3.4 Effluent control

As the solutions are acidic and contain nickel, effluent must be treated before disposal.

14.4 NON-STICK PTFE COATINGS

Non-stick coatings are used extensively for domestic hollow ware and baking utensils. Originally applied as coatings directly on to aluminium, but being soft, they were easily damaged by metal utensils, requiring use of plastic or wooden spoons and spatulas.

In the early days, the problem was to obtain good adhesion to the aluminium. This was overcome by etching the aluminium to form deep pits by chloride-containing etches such as 5% hydrochloric acid at 60°C, to form a key with the PTFE. The coatings, known under the trade names of Fluon, Tefal, Teflon etc., are applied as a slurry of fine particles with water, which are then followed by drying and melting at about 400°C.

As an alternative to etching, sand blasting has been used to provide a key. A prime coat consisting of PTFE with chromic or chromic and phosphoric acids has been used, which is applied slightly under-cured so that the top coat will adhere firmly to it on final stoving. These acid-containing prime coats are not suitable for cooking utensils on account of possible toxicity.

Because of the poor wear resistance of PTFE coatings on their own, increased abrasion resistance has been provided by incorporating into the coating hard particles. Aluminium powder has been flame sprayed on to the aluminium before coating with PTFE. A second method has been to impinge stainless steel particles on to the surface, a method which is used in engineering applications requiring a low-friction and hard wearing bearing surface.

A method currently being used is to apply a stove-enamel frit which forms after firing discrete particles which promote adhesion and provide wear resistance to the subsequently applied PTFE coating.

14.5 FLAME SPRAYING

This is the general term for the application of various substances in the semi-molten condition on to a metal surface, which includes metals, ceramics, etc. It is carried out by projecting fine particles through a hot flame on to a surface on which they are condensed.

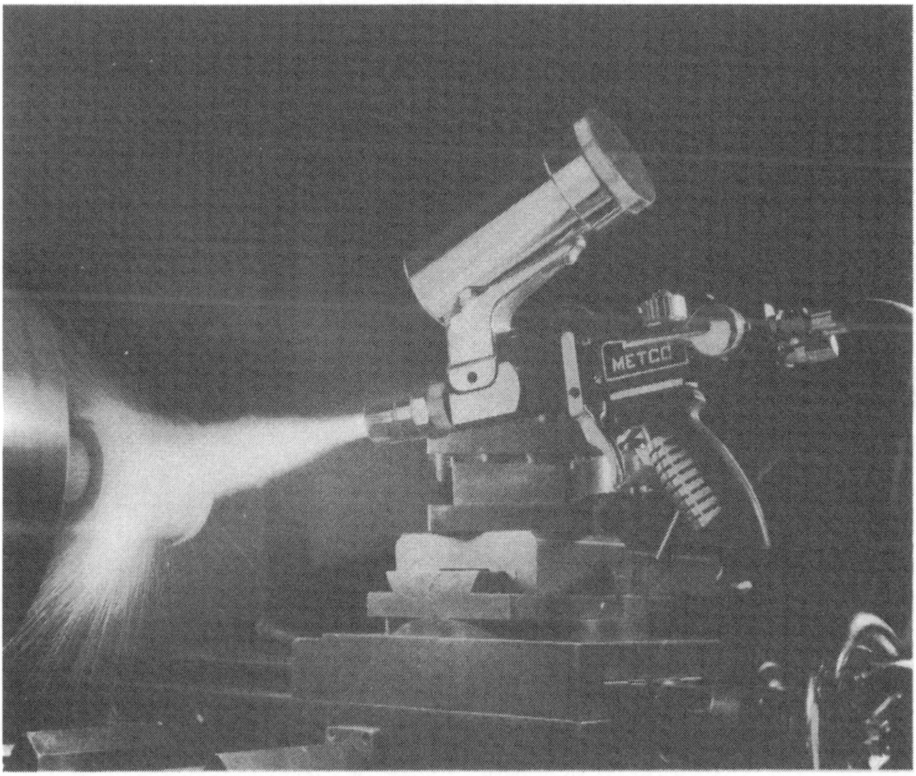

FIGURE 14.1. Metal spraying from powder — Metco Ltd

Metal spraying is carried out by projecting fine drops of molten metal on to another metal to form an adherent coating. It is a process usually carried out to coat steel with aluminium or zinc to provide a galvanic protective layer against corrosion.

Mostly, the aluminium used for spraying is supplied in wire form where it is fed to the gun, melted, and expelled as molten droplets by compressed air or a non-oxidising gas such as nitrogen.

Metal spraying of pure aluminium on to components made from copper-containing aluminium alloys is a successful method of providing protection to these alloys. An early example of this was a crane jib in 2014 alloy which was protected in this way, and metal spraying with aluminium containing 1% zinc has been found to be successful in protecting 2014 aluminium from

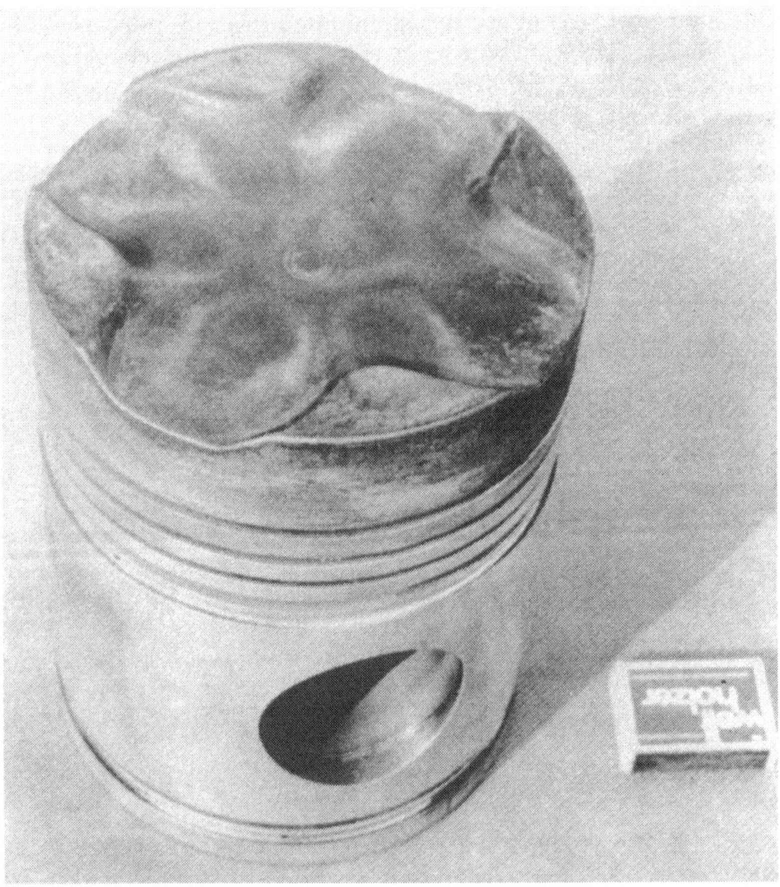

FIGURE 14.2. Cast aluminium piston for railway diesel engine. The head was plasma flame sprayed with nickel aluminide and aluminium to prevent heat damage and provide corrosion resistance — Metco Ltd

FIGURE 14.3. Helicopter deck on stern of USS destroyer *Truitt*, made from aluminium, was sprayed with nickel aluminide and aluminium for corrosion resistance and to provide a non-skid surface — Metco Ltd

stress corrosion, as well as surface corrosion or exfoliation. It is useful for the protection of weld areas, particularly for structures in high-strength alloys where welds can suffer from stress–corrosion in corrosive environments.

Heat treatable aluminium castings containing copper have been sprayed with 99.5% aluminium for protection.

Flame spraying of cermets, which are mixtures of ceramics and metals, has been used to build a hard, thick and heat resistant coating on aluminium to provide both resistance to wear and to high temperatures owing to its thermal insulation properties. These properties have been used to advantage as liners for aluminium cylinder blocks.

Defence Standard 03–3 "Protection of Aluminium Alloys by Sprayed Metal Coatings" specifies the requirements of these coatings, which include zinc and aluminium-zinc alloys as well as pure aluminium. The Standard describes the pretreatment required, which includes cleaning followed by abrasive blasting with aluminium. The average coating thickness is required to be between 100 and 200 μm, and local thickness not less than 75 μm.

14.6 ELECTROPHORETIC LACQUER

An interesting process for lacquering metals is marketed by Albright & Wilson Ltd. under the name Albrifin. This is a water-based clear lacquer which is deposited electrophoretically on to metals, and before stoving the lacquer coating is dyed in the same way as are anodised coatings. Unlike anodising, the colours are not dependent on coating thickness, and a range of colours is available. Producing colour in this way avoids the need to have different coloured lacquers available, thus saving space and cost.

The lacquers can be applied to a variety of finishes on aluminium. Although the stoved coatings are perhaps not as hard or durable as anodic coatings, the lacquer process is not affected by or dependent on the purity or metallurgical structure of the aluminium. It can be applied to either etched or mechanically polished aluminium without the need for chemical polishing.

It would seem to have particular interest in the processing of aluminium castings which, because of their structure and their high silicon content, are not suitable for decorative anodising or for chemical brightening processes.

14.7 TEMPORARY PROTECTION

Aluminium, whether precoated or not, is vulnerable to damage from handling or mechanical damage before it is finally assembled or manufactured. Ways of protecting the surface by applying a film or coating are available which are able to provide protection, either physically or chemically, and if need be, can be removed before or on completion.

The simplest ways are coatings of oil, grease or wax for protection against corrosion after manufacture and before final and permanent coatings are applied. Lanolin-based grease is used for this type of protection, whilst waxes have been used on anodised coatings to prevent chemical damage from alkaline wet-mortar splashing. Waxes, in these cases, can be left to weather away unobtrusively. Air-drying and non-yellowing lacquers based on butyrates and methacrylates are also used in this way.

Mechanical protection can be provided by coating sheet with special adhesive backed papers or vinyl plastic film. Whilst paper can be used for protection during limited forming operations, the greater flexibility of the plastic coatings, although more expensive, are more suitable.

Vinyl coatings can also be supplied as a thin film from a liquid by spraying or other means. The coatings have limited adhesion so that they can be peeled off when required.

The coatings should have the following properties:

1. Flexible, if post-forming required
2. Low-tack adhesion
3. Should not harden on ageing or exposure to UV radiation
4. Thick and strong enough to provide protection.

The adhesives used for tapes and film should be not too strong, and should be easily removable after being in place for a long interval. The adhesive, on the other hand, must have good adhesion to the tapes to avoid residues of adhesive remaining on the aluminium when the tape is removed.

Some adhesives used with tape, as well as the tape itself, can harden on ageing and may prove difficult to remove. It is for this reason that protective tapes are not often used on painted aluminium, as the tape removal can have unwelcome effects on the paint adhesion.

A problem with solution vinyl coatings when used as temporary protectives is that if they are applied too thinly, they do not separate in one piece on removal. They also tend to harden on ageing.

Paper or vinyl films can be applied to coiled aluminium by special applicators, or they can be applied to individual flat sheets. Tape can also be applied to particular flat surfaces of extrusions that may need protection from damage. This is often carried out on architectural sections, perhaps fabricated into assembled window frames, before they are erected on a building.

14.8 HEAT SHRINK PLASTICS

Certain formulations of plastics under the action of heat are caused to shrink. This effect has been used to coat aluminium tube or section. The plastic is supplied as a flexible tube which can be passed over the aluminium tube or simple section and heat is applied by passing the length through a flame to heat the plastic but not, of course, to char it. The plastic contracts and forms a tight coating round the aluminium. The plastic can be coloured and coatings about 3 millimetres thick can be applied. They have been used on tube used for handles of garden tools.

Appendix I

There are a number of books and sources of knowledge in which more detailed technical and commercial information can be obtained on the finishing of aluminium, the following organisations being of immediate interest.

Some of the following organisations are able to offer technical assistance as well as being able to offer advice.

Aluminium Federation Ltd, Broadway House, Calthorpe Road, Fiveways, Birmingham B15 1TN
BNF Metals Technology Centre, Denchworth Road, Wantage, Oxfordshire OX12 9BJ
Alcan International Ltd., Southern Road, Banbury, Oxfordshire
Alcoa of Great Britain Ltd., P.O. Box 15, Droitwich, Worcestershire WR9 7BG
British Alcan Aluminium plc, Chalfont Park, Gerrards Cross, Buckinghamshire
Paint Research Association, Waldegrove Road, Teddington, Middlesex.

Many universities are able to offer help as well as Paint Manufacturing companies and Chemical Supply Houses.

The Institute of Metal Finishing, Exeter House, 48 Holloway Head, Birmingham B1 1NQ is able to offer assistance through their members. It publishes quarterly the *Transactions of the Institute of Metal Finishing* which contains papers of technical content, as well as holding an annual Technical Conference, backed by many local meetings throughout the year.

The following monthly publications include much useful technical and commercial information:

Finishing, Wheatland Journals Ltd., 177 Hagden Lane, Watford WD1 8LW
Product Finishing, Sawell Publications Ltd., 127 Standstead Road, Forest Hill, London SE23
Metal Finishing, Metals & Plastics Publishers Inc., One University Plaza, Hackensack, New Jersey 076601
Metal Finishing Abstracts (bi-monthly), Finishing Publications, 17 Cramner Road, Hampton Hill, Middlesex.

There are many books which specialise in various aspects of metal finishing and give more detailed information on theoretical aspects. These can be purchased, or can be consulted from many of the larger libraries. They can usually be borrowed from British Library Lending Division (Boston Spar 843434). One book specific to aluminium is:

The Surface Treatment of Aluminium and its Alloys, S. Wernick and R. Pinner; Fourth Edition 1972, 2 volumes, 1274 pp. This edition is one of the most comprehensive books on finishing and contains many references to all aspects of finishing. A later edition has now been published.

Some specialised books on various subjects are as follows. They are not confined only to aluminium but deal with other metals as well.

Conversion Coatings

Electrolytic & Chemical Conversion Coatings, T. Biestek and S. Weber, Portcullis Press Ltd., 432 pp, 1976.

Electro-Plating

Canning Handbook on Electro-Plating, W. Canning Materials Ltd., 1020 pp, 1978.

Mechanical Treatments

Mechanical Polishing, Buskart, Silman and Draper, Portcullis Press Ltd., 267 pp, 1960.

Organic Coatings

Electro Painting, R. L. Yeates, Portcullis Press Ltd., 278 pp, 1970.
Fundamentals of Powder Coating, E. P. Miller and D. D. Taft, Society of Manufacturing Engineers, 287 pp, 1974.
Industrial Paint Application, W. H. Tatton and E. W. Drew, Newnes-Butterworths, 272 pp, 1971.

Testing

Corrosion Testing for Metal Finishing, V. E. Carter Institute of Metal Finishing.
Corrosion Prevention Directory, Department of Industry, HMSO, 136 pp. This book lists sources of information, has a directory of books, as well as giving useful information on the protection of metals.

Vitreous Enamel

Bulletin AL–2A *Recommended Processing Methods for Porcelain Enamel on Aluminum Alloys*, Porcelain Enamel Institute Inc., USA.

General

Principles of Metal Surface Treatment and Protection, D. R. Gabe, Pergamon Press, 181 pp, 1972.

Appendix II

The following standards relate to finishes that are applied to aluminium.

BRITISH STANDARDS

BS 1244	Electroplated Coatings of Nickel and Chromium
BS 1344	Methods of Testing Vitreous Enamel Finishes
BS 3900	Methods of Tests for Paints
BS 4800	Specification for Paint Colours for Building Purposes
BS 4842	Stoving Organic Finishes on Aluminium Extrusions and Preformed Sections for External Architectural Application
BS 5252	Framework for Colour Co-ordination for Building Purposes
BS 5466	Methods of Corrosion Testing of Metallic Coatings
BS 6496	Specification for Powder Organic Coatings for Application and Stoving of Aluminium Alloy Extrusions, Sheet and Preformed Sections for External Architectural Purposes and the Finish on Aluminium Alloy and Extrusions, Sheet and Preformed Sections Coated with Powder Organic Coatings

MINISTRY OF DEFENCE

DEF 1033	Standard Method of Testing Paint Varnish, Lacquer and Related Products
DEF STAN 03–2	Cleaning and Preparation of Metal Surfaces
DEF STAN 03–3	Protection of Aluminium Alloys by Sprayed Metal Coatings

In addition, and sometimes parallel to, there are a number of International Standards (ISO) covering coatings applied to aluminium.

In the United States, a series of standards exist, some of the relevant titles are as follows:

AMERICAN SOCIETY FOR TESTING AND MATERIALS

B253–68	Recommended Practice for Preparation of and Electroplating on Aluminum Alloy by the Zincate Process
B322–68	Recommended Practice for Cleaning of Metals Prior to Electroplating
B449–67	Recommended Practice for Chromate Treatments of Aluminum
D1730–67	Recommended Practice for Preparation of Aluminium and Aluminum Alloy Surfaces for Painting

In addition, there are a number of ASTM specifications for testing paint and other coatings which can be applicable to aluminium.

Various US Military Specifications exist and are as follows:

AERONAUTICAL MATERIAL SPECIFICATIONS

AMS 2420	Plating aluminum for solderability (zincate process)
AMS 2436	Flame deposition of aluminum oxide coating
AMS 2473A	Chemical treatment for aluminum alloys—general purpose
AMS 2474	Chemical treatment for aluminum alloys— low electrical resistivity
AMS 2505A	Aluminum paint finishing —application of zinc chromate and aluminum paint on parts

MILITARY SPECIFICATIONS

MIL-C-5541	Chemical films for aluminum and aluminum alloys (for paint adhesion or corrosion resistance)
MIL-P-6808A	Application of zinc chromate primer to aircraft
MIL-T-704C	Treatment and painting for construction and engineering equipment. (Various metals including aluminum)
MIL-S-5002	Surface treatments for metal and metal parts in aircraft
MIL-P-14504A	Primer coating pretreatment (single-pack polyvinyl butyral resin)
MIL-C-15328A	Wash primer pretreatment (phosphoric acid type wash primer)
MIL-P-15930A	Vinyl-zinc chromate primer (to follow MIL-C-15328A)
MIL-A-16994B	Porcelain enamel coated sheet (6061 sheet)
MIL-F-7179A	Finishes and coatings. General specifications for finishing of aircraft and aircraft parts

Index

Made in the USA
Monee, IL
13 September 2026

62977002R10090